SpringerBriefs in Fire

Series editor

James A. Milke, University of Maryland, College Park, USA

More information about this series at http://www.springer.com/series/10476

Andrew F. Blum · R. Thomas Long Jr.

Fire Hazard Assessment of Lithium Ion Battery Energy Storage Systems

Springer

Andrew F. Blum
Exponent, Inc.
Bowie, MD
USA

R. Thomas Long Jr.
Exponent, Inc.
Bowie, MD
USA

ISSN 2193-6595 ISSN 2193-6609 (electronic)
SpringerBriefs in Fire
ISBN 978-1-4939-6555-7 ISBN 978-1-4939-6556-4 (eBook)
DOI 10.1007/978-1-4939-6556-4

Library of Congress Control Number: 2016946952

Printed on acid-free paper

This Springer imprint is published by Springer Nature
The registered company is Springer Science+Business Media LLC New York

Foreword

At the request of the Fire Protection Research Foundation (FPRF), exponent performed a fire hazard assessment of lithium-ion (Li-ion) batteries used in energy storage systems (ESSs). This book summarizes a literature review and gap analysis related to Li-ion battery ESSs, as well as full-scale fire testing of 100-kWh Li-ion battery ESSs. The scope of services performed during this literature review and testing program may not adequately address the needs of other users of this book, and any reuse of this book or the findings, conclusions, or recommendations presented herein are at the sole risk of the user.

The full-scale Li-ion battery ESS test strategy, ignition protocols, and any recommendations made are strictly limited to the test conditions included and detailed in this book. The combined effects (including, but not limited to) of different battery types, ESS types, ESS size/battery capacity, internal or external ESS/battery damage, battery energy density and design, state of charge, and cell chemistry are yet to be fully understood and may not be inferred from these test results alone.

The findings formulated in this review are based on the observations and information available at the time of writing. The findings presented herein are made to a reasonable degree of scientific and engineering certainty. If new data become available or there are perceived omissions or misstatements in this book, we ask that they be brought to our attention as soon as possible so that we have the opportunity to fully address them.

Bowie, MD, USA

Andrew F. Blum
R. Thomas Long Jr.

Preface

In an effort to provide guidance to standards developers, authorities having jurisdiction (AHJs), emergency responders, and the energy storage system (ESS) industry, exponent, in conjunction with FPRF, the Project Technical Panel, and industry sponsors, performed a fire hazard assessment of Li-ion battery ESSs. Currently, these entities do not have a clear direction regarding the fire hazards of ESS installations and have few, if any, technical studies, reports, or scientific literature to rely upon when making decisions regarding the safe installation of these systems. This book summarizes a literature review and gap analysis related to Li-ion battery ESSs, as well as full-scale fire testing of a 100-kWh Li-ion battery ESS.

The scope of work included, but was not limited to, the following four primary tasks:

1. A literature review and gap analysis related to Li-ion battery ESSs;
2. Development of a detailed full-scale fire testing plan to perform an assessment of Li-ion battery ESS fire hazards;
3. Witnessing the implementation of the fire test plan through full-scale fire testing; and
4. A report of final results and a fire hazard assessment.

The overall project research objective was to develop a technical basis through a fire hazard assessment of Li-ion battery ESSs. This project is the first phase of an overall initiative with the goal to develop safe installation practices, fire protection guidance, and appropriate emergency response tactics for Li-ion battery ESSs based on the literature review and full-scale test results, as applicable. This project did not include an analysis or testing of fire detection systems, fire suppression systems, or emergency response tactics related to Li-ion battery ESS fire scenarios. A full listing of project observations/key findings as they relate to ESS fire hazards is provided in Chap. 7 of this book.

Bowie, MD, USA

Andrew F. Blum
R. Thomas Long Jr.

Acknowledgements

The authors would like to thank Tesla for donating the Powerpacks and for their significant efforts during this project.

The authors further thank Kathleen Almand, Executive Director, FPRF; Daniel Gorham, Research Project Manager; and everyone on the FPRF Advisory Panel.

Contents

Acronyms and Abbreviations

Ah	Ampere hours
AHJ	Authority having jurisdiction
APS	Arizona Public Service Company
BATSO	Battery Safety Organization
BMS	Battery management system
CH_4	Methane
Cl_2	Chlorine
CO	Carbon monoxide
CO_2	Carbon dioxide
DC	Direct current
DOE	Department of Energy
DOT	Department of Transportation
EDV	Electric drive vehicle
EES	Electrical energy storage
ELC	Equivalent lithium content
ESS	Energy storage system
FMEA	Failure modes and effects analysis
FPRF	Fire Protection Research Foundation
ft	Feet
g	Grams
HCl	Hydrogen chloride
HCN	Hydrogen cyanide
HECO	Hawaiian Electric Company
HF	Hydrogen fluoride
HFD	Honolulu Fire Department
HFG	Heat flux gauge
HRR	Heat release rate
IBC	International Building Code
ICC	International Code Council
ICE	Internal combustion engine

IEC	International Electrotechnical Commission
IEEE	Institute of Electrical and Electronics Engineers
IFC	International Fire Code
in	Inch
IRC	International Residential Code
JSA	Japanese Standards Association
kg	Kilograms
kWh	Kilowatt hours
LER	Light electric rail
Li-ion	Lithium ion
lpm	Liters per minute
mph	Miles per hour
MW	Megawatt
NEC	NFPA 70, *National Electrical Code*
NEMA	National Electrical Manufacturers Association
NFIRS	National Fire Incident Reporting System
NFPA	National Fire Protection Association
NiCad	Nickel cadmium
NO_x	Nitrogen oxides
OSHA	Occupational Safety and Health Administration
PF_5	Phosphorus pentafluoride
POF_3	Phosphoryl fluoride
ppm	Parts per million
psi	Pounds per square inch
PVES	Photovoltaic energy systems
SAE	Society of Automotive Engineers
SCBA	Self-contained breathing apparatus
SDS	Safety data sheet
SOC	State of charge
TC	Thermocouple
UL	Underwriters Laboratories
UN	United Nations
UPS	Uninterrupted power supplies
VOC	Volatile organic compound
Wh	Watt hours

List of Figures

List of Tables

Chapter 1
Background

1.1 Project History

Energy storage is emerging as an integral component of a resilient and efficient electrical grid through a diverse array of potential applications. It is anticipated that the evolution of the electrical grid will result in a greater need for services best provided by energy storage systems (ESSs). It is expected that the increase in demand for these systems will further drive energy storage research to produce systems with greater efficiency at a lower cost, which will lead to an influx of energy storage deployment across the country. To enable the success of these deployments, the hazards of these systems, namely the fire hazard of the ESS, must be understood.[1]

In recent years, there has been a marked increase in the deployment of lithium ion (Li-ion) batteries in ESSs. Many ESSs are being deployed in both high-rise structures and single- and multi-family residences. Local authorities having jurisdiction (AHJs) along with ESS integrators and installers do not have a clear direction regarding the fire hazards of these installations. A recognized fire hazard assessment available to standards developers, AHJs, emergency responders, and industry will provide guidance with a technical basis on the evaluation and safe installation of these systems.

1.2 Research Objectives and Project Scope

The overall project research objective was to develop a technical basis through a fire hazard assessment of Li-ion ESSs. This project is part of an overall initiative with the goal to develop safe installation practices, fire protection guidance, and

---

[1]Energy Storage Safety Strategic Plan, U.S. Department of Energy, December 2014.

A.F. Blum and R.T Long Jr., *Fire Hazard Assessment of Lithium Ion Battery Energy Storage Systems*, SpringerBriefs in Fire,
DOI 10.1007/978-1-4939-6556-4_1

appropriate emergency response tactics for ESSs. This project did not include an analysis or testing of fire detection systems, fire suppression systems, or emergency response tactics related to Li-ion battery ESS fire scenarios.

The scope of work included, but was not limited to, the following four primary tasks:

1. A literature review and gap analysis related to Li-ion battery ESSs;
2. Development of a detailed full-scale fire testing plan to perform an assessment of Li-ion battery ESS fire hazards;
3. Witnessing the implementation of the fire test plan through full-scale fire testing; and
4. A report of final results and a fire hazard assessment.

A more detailed description of the tasks Exponent performed to fulfill the project objectives is provided below.

1.2.1 Literature Review and Gap Analysis

Exponent collected, reviewed, and summarized available literature related to Li-ion battery ESSs, including the Department of Energy (DOE) Safety Roadmap, relevant codes and standards, incident reports, related test plans, and previous fire testing/research. The literature review also identified existing gaps in the information currently available and the practices utilized in the deployment of Li-ion ESSs, if any.

1.2.2 Fire Test Plan

Exponent, in conjunction with the Project Technical Panel, developed a detailed test plan to provide an assessment of fire hazards posed by Li-ion ESSs. Li-ion ESSs with an approximate capacity of 100 kilowatt hours (kWh) designed for use in commercial applications were tested.

1.2.3 Witness of Fire Testing

Exponent witnessed the full-scale fire testing at the manufacturer's testing site and summarized the test observations and data provided to Exponent.

1.2.4 Final Report

Exponent collected and summarized the results of the above tasks in a formal research engineering report, including:

1. An overview of the project work to date;
2. A summary of the full-scale fire tests;
3. A fire hazard assessment; and
4. Identification of future potential research.

Chapter 2
Literature Review and Gap Analysis

Exponent collected, reviewed, and summarized available literature related to ESSs and Li-ion batteries. The literature review provides an overview of energy storage (Sect. 2.1), commercial and residential ESSs (Sect. 2.2), a brief summary of Li-ion technology (Sect. 2.3), codes and standards related to ESSs (Sect. 2.4), fire incidents involving ESSs (Sect. 2.5), large format Li-ion battery fires (Sect. 2.6), and a gap analysis (Sect. 2.7).

2.1 Energy Storage Overview

An ESS provides a means to store energy for later use to supply the utility grid or local grids.[1] An ESS may utilize any of the following technologies:

1. **Electrochemical**. Consists of a secondary battery, electrochemical capacitor, flow battery, or hybrid battery-capacitor system that stores energy and any associated controls or devices that can provide electric energy upon demand.
2. **Chemical**. Consists of hydrogen supply equipment or other fuel supply equipment combined with a fuel cell power system or generator to convert the fuel to electrical energy.
3. **Mechanical**. Consists of a mechanical means to store energy, such as through compressed air, pumped water, or fly wheel technologies and associated controls and systems, which can be used to run an electric generator to provide electric energy upon demand.
4. **Thermal**. Consists of a system that uses heated fluids, such as air, as a means to store energy along with associated controls and systems, which can be used to run an electric generator to provide electrical energy upon demand.

[1]UL 9540, *Outline of Investigation for Energy Storage Systems and Equipment*, Issue Number 1, June 30, 2014.

A.F. Blum and R.T Long Jr., *Fire Hazard Assessment of Lithium Ion Battery Energy Storage Systems*, SpringerBriefs in Fire,
DOI 10.1007/978-1-4939-6556-4_2

This book focuses on Li-ion battery ESSs for commercial and residential installations, which are an electrochemical technology.

An ESS allows for the balance of supply and demand of electrical energy, utilizing stored energy during "peak demand" times and storing energy during times of "low demand." An example of a common ESS is pumped-storage hydroelectricity (pumped hydro). Pumped hydro stores large quantities of water in elevated reservoirs by utilizing excess electricity at times of low demand to pump water into the reservoirs. The facilities then release the water, which passes through turbine generators and converts the stored potential energy to electricity when electrical demand peaks [1].

Recently, a more common solution is the storage of energy in a battery. Batteries have historically been of limited use in large scale electric power systems due to their relatively small capacity and high cost. However, newer battery technologies have been developed that can provide significant utility scale capabilities [1]. In addition to utility scale applications, smaller commercial and residential ESSs utilizing batteries are also becoming more prevalent.

2.2 Commercial and Residential ESS Overview

The most common commercial and residential ESSs are electrochemical systems utilizing batteries. Currently, there are many different battery chemistries (e.g., lead acid, sodium sulfur, lithium iron phosphate, Li-ion) utilized in ESSs deployed in North America; however, Li-ion is the most popular[2] and will likely continue to grow in popularity with the planned release of new ESS products in the coming years.

Residential ESSs are typically sized between 1 and 10 kWh,[3, 4, 5] and standalone commercial systems can be much larger (20–100 kWh), modular, and interconnected to produce even greater capacity. The systems can vary in voltage depending on the design of the batteries, the ESS power management systems, and the manufacturer. Current products installed in the market have voltages as low as 48 V and as high as 1000 V DC. ESSs typically work by storing power collected from the grid, a solar installation, wind installation, or other source during a low demand time (typically during the day) and then using the stored energy during peak hours (typically in the mornings and evenings), as illustrated in Fig. 2.1 (see footnote 3, 4).

---

[2]Energy Storage Safety Strategic Plan, U.S. Department of Energy, December 2014.

[3]http://www.samsungsdi.com/ess/overview.

[4]http://www.teslamotors.com/powerwall.

[5]http://www.aquionenergy.com/energy-storage-battery.

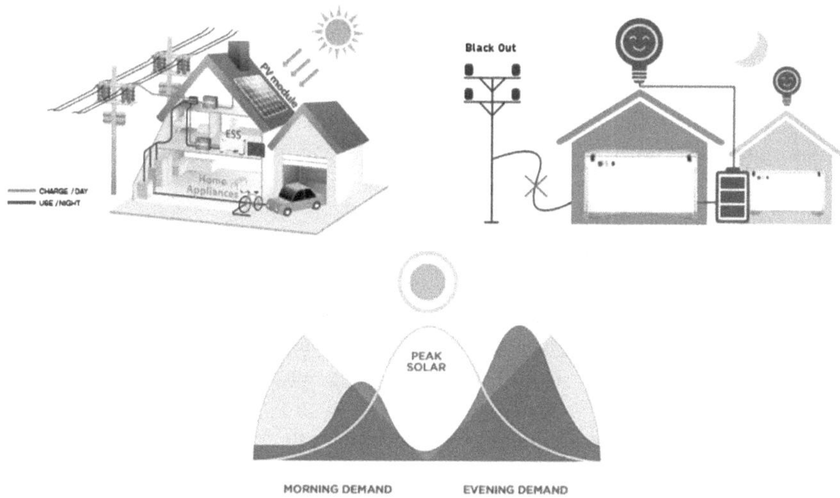

Fig. 2.1 Illustration of energy storage during off peak hours (*top left*); use of energy storage during peak hours or power interruptions (*top right*); and the typical energy consumption curve (*bottom*) (http://www.samsungsdi.com/ess/residential-commercial-solution)

The ESS typically consists of the batteries, a mounting frame or shelf for the batteries, a cooling system (i.e., fan, radiator, and hoses), power electronics, and an enclosure (the outer cover or cabinet) that these components are stored within. A residential ESS can be installed inside a residence or building, typically within the garage or attic, or installed on the exterior of the structure. A commercial ESS can be installed outside along a property line, next to a building, or inside a shipping container.

Pumped hydro remains one of the oldest and most mature energy storage technologies, having been utilized safely since the 1800s. Its hazards are well known and defined. Battery ESSs, however, are much earlier in their development and deployment cycle and, given recent trends, have not reached the full extent of their deployed capacity.[6] The hazards associated with these systems are not well known and are less defined than other traditional ESS technologies, such as pumped hydro. When discussing ESSs in the remainder of the book, Exponent is referring to Li-ion battery ESSs for use in commercial applications.

[6]Energy Storage Safety Strategic Plan, U.S. Department of Energy, December 2014.

2.3 Li-ion Battery Overview

Li-ion battery cells are in wide consumer use today. As this technology has evolved and the energy densities have increased, the use of this technology has been applied across many consumer products, including the energy storage industry. Li-ion battery cells arranged in large format Li-ion battery packs are being used to power ESSs. As ESSs enter the United States consumer marketplace, there is an expectation of a steep increase in the number and size of battery packs in storage and use. Recent studies by the National Fire Protection Association (NFPA) Fire Protection Research Foundation (FPRF) [2–5] highlight the potential hazards of Li-ion battery cells and large format packs during the life cycle of storage, distribution, and use in products. An overview of the Li-ion technology and its failure modes is also included. A brief summary of Li-ion technology is provided here.

Li-ion has become the dominant rechargeable battery chemistry for consumer electronic devices and is poised to become commonplace for industrial, transportation, and energy storage applications. This chemistry is different from previously popular rechargeable battery chemistries (e.g., nickel metal hydride, nickel cadmium, and lead acid) in a number of ways. From a technological standpoint, because of high energy density, Li-ion technology is an effective battery type to use in ESSs. From a safety and fire protection standpoint, a high energy density coupled with a flammable organic, rather than aqueous, electrolyte has created a number of new challenges with regard to the design of batteries containing Li-ion cells, and with regard to fire suppression.

2.3.1 Anatomy of a Li-ion Cell

The term "Li-ion" refers to an entire family of battery chemistries. It is beyond the scope of this book to describe all of the chemistries used in commercial Li-ion batteries. In addition, Li-ion battery chemistry is an active area of research and new materials are constantly being developed. Additional detailed information with regard to Li-ion batteries is available in a number of references[7, 8] and a large volume of research publications and conference proceedings on the subject.

In the most basic sense, the term "Li-ion battery" refers to a battery where the negative electrode (anode) and positive electrode (cathode) materials serve as a host for the lithium ion (Li+). Lithium ions move from the anode to the cathode during discharge and are intercalated (inserted into voids) in the crystallographic structure of the cathode. The ions reverse direction during charging, as shown in Fig. 2.2. Since lithium ions are intercalated into host materials during charge or discharge,

[7] *Linden's Handbook of Batteries*, 4th Edition, Thomas B. Reddy (ed), McGraw Hill, NY, 2011.

[8] *Advances in Lithium-Ion Batteries*, WA van Schalkwijk and B Scrosati (eds), Kluwer Academic/Plenum Publishers, NY, 2002.

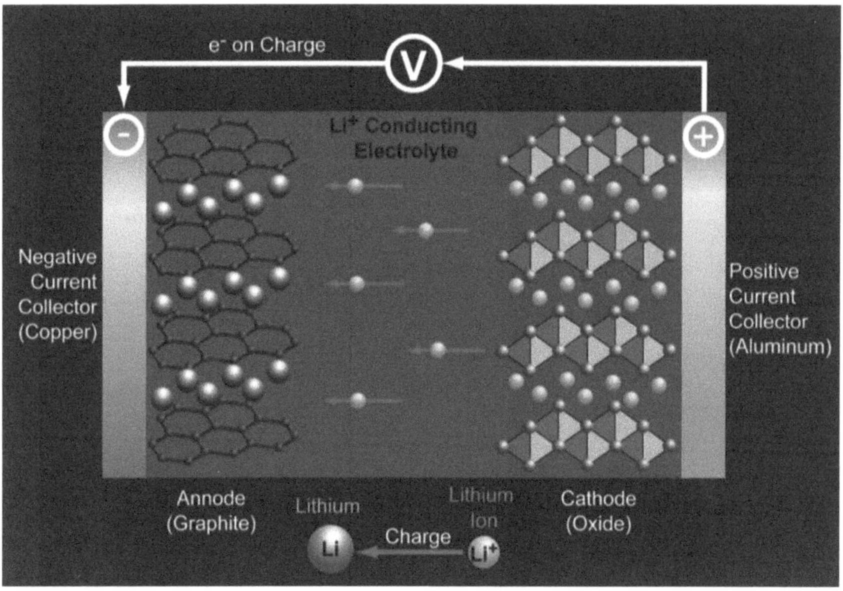

Fig. 2.2 Li-ion cell operation: during charging, lithium ions intercalate into the anode, the reverse occurs during discharge

there is no free lithium metal within a Li-ion cell,[9, 10] thus, if a cell ignites due to external flame impingement or an internal fault, metal fire suppression techniques are not appropriate for controlling the fire.

In a Li-ion cell, alternating layers of anodes and cathodes are separated by a porous film (separator). An electrolyte composed of an organic solvent and dissolved lithium salt provides the media for Li-ion transport. A cell can be constructed by stacking alternating layers of electrodes (typical for high-rate capability prismatic cells), or by winding long strips of electrodes into a "jelly roll" configuration typical for cylindrical cells, as shown in Fig. 2.3. Electrode stacks or rolls can be inserted into hard cases that are sealed with gaskets (most commercial cylindrical cells), laser-welded hard cases, or enclosed in foil pouches with heat-sealed seams (commonly referred to as Li-ion polymer cells[11]), as shown in Fig. 2.4. A variety of safety mechanisms might also be included in the mechanical

[9]Under certain abuse conditions, lithium metal in very small quantities can plate onto anode surfaces. However, this should not have any appreciable effect on the fire behavior of the cell.

[10]There has been some discussion about the possibility of "thermite-style" reactions occurring within cells. See the NFPA FPRF report titled, "Lithium-Ion Batteries Hazard and Use Assessment," for an in-depth analysis.

[11]The term "lithium polymer" has been previously used to describe lithium metal rechargeable cells that utilized a polymer-based electrolyte. Lithium polymer is now used to describe a wide range of Li-ion cells enclosed in soft pouches with electrolyte that may or may not be polymer based.

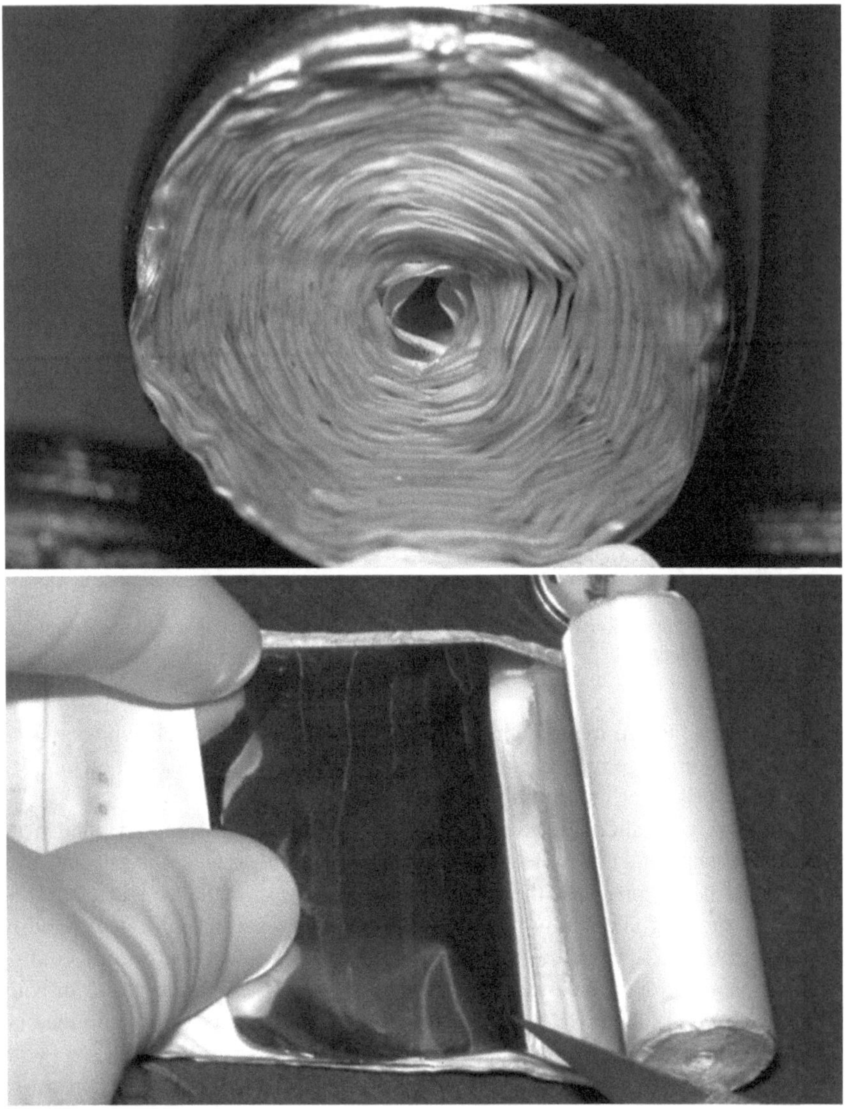

Fig. 2.3 Base of a cylindrical Li-ion cell showing wound structure (*top*); Cell being unwound revealing multiple layers: separator is *white*, aluminum current collector (part of cathode) appears shiny (*bottom*)

design of a cell, such as charge interrupt devices and positive temperature coefficient switches.[12, 13]

[12]For a more detailed discussion of Li-ion cells see: Dahn and Ehrlich [6].

[13]For a review of various safety mechanisms that can be applied to Li-ion cells see: Balakrishnan et al. [7].

Fig. 2.4 Example of 18,650 cylindrical cells; these are the most common consumer electronics Li-ion cell form factor (*top*); hard case prismatic cell (*bottom left*); and soft pouch polymer cell (*bottom right*)

2.3.2 Li-ion Cell Characteristics and Hazards

The electrolyte within a typical Li-ion cell includes a volatile hydrocarbon-based liquid and a dissolved lithium salt (which is a source of lithium ions), such as lithium hexofluorophosphate. Battery cells are hermetically sealed to prevent moisture in the air from degrading the cells. Li-ion cells are not vented to the atmosphere like lead acid batteries, therefore, under normal usage conditions, they do not exhaust vapors. In normal usage, cell electrolyte should not be encountered by anyone handling a Li-ion battery, making the risk of a spill of electrolyte from any commercial Li-ion battery pack very remote. Furthermore, in most commercial cells, the electrolyte is largely absorbed in electrodes, such that there is no free or "spillable" electrolyte within individual sealed cells. In those instances, severe mechanical damage (e.g., severe crushing) can cause a small fraction of total electrolyte quantity to leak out of a single cell; however, any released electrolyte is likely to evaporate rapidly.

Li-ion cells are sealed units, and thus under normal usage conditions, venting of electrolyte should not occur. If subjected to abnormal heating or other abuse conditions, electrolyte and electrolyte decomposition products can vaporize and be vented from cells. Accumulation of liquid electrolyte is unlikely in the case of abnormal heating. Vented electrolyte is flammable, and may ignite on contact with a competent ignition source, such as an open flame, spark, or a sufficiently heated surface. Vented electrolyte may also ignite on contact with cells undergoing a thermal runaway reaction. Cell vent gas composition will depend upon a number of factors, including cell composition, cell state of charge, and the cause of cell venting. Vent gases may include volatile organic compounds (VOCs, such as alkyl-carbonates, methane, ethylene, and ethane), hydrogen gas, carbon dioxide, carbon monoxide, soot, and particulates containing oxides of nickel, aluminum, lithium, copper, and cobalt. Additionally, phosphorus pentafluoride (PF_5), phosphoryl fluoride (POF_3), and hydrogen fluoride (HF) vapors may form. Vented gases may irritate the eyes, skin, and throat. Cell vent gases are typically hot and upon exit from a cell, can exceed 600 °C (1112 °F). Contact with hot gases can cause thermal burns.[14]

2.3.3 *Li-ion Battery Design*

A Li-ion battery is made from multiple individual cells packaged together with their associated control system and protection electronics. By connecting cells in parallel, designers increase pack capacity. By connecting cells in series, designers increase pack voltage. Thus, most battery packs will be labeled with a nominal voltage that can be used to infer the number of series elements and, along with total battery pack energy (in watt hours [Wh]), can be used to determine the capacity (in ampere hours [Ah]) of each series element (size of individual cells or the number of cells connected in parallel). A Li-ion battery, despite conformance to a number of safety standards, may pose a significant high voltage and electrocution risk if it has been significantly damaged. Since Li-ion cells are not cycled to 0 V, a Li-ion battery pack, even in a normally discharged condition, is likely to contain substantial electrical charge. Cutting into a normally discharged battery pack can cause sparking or create electrocution hazards.

For large format battery packs, cells may be connected together (in series and/or in parallel) in modules. The modules may then be connected in series or in parallel to form full battery packs. Modules are used to facilitate readily changed

[14]Lithium-Ion Battery Emergency Response Guide, Tesla Energy Products, September 2015, Revision 02.

configurations and easy replacement of faulty portions of large battery packs. Thus, large format battery pack architecture can be complex.

ESS batteries typically utilize many individual cells comprised into modules, which are assembled to form a large format battery pack. Large format battery packs typically contain an active safeguarding system to monitor electrical current, voltage, and temperature of the cells to optimize pack performance and mitigate potential failures, including fire. Numerous standards and protocols are available for these packs, including documents created by Underwriters Laboratories (UL), Institute of Electrical and Electronics Engineers (IEEE), National Electrical Manufacturers Association (NEMA), Society of Automotive Engineers (SAE), International Electrotechnical Commission (IEC), United Nations (UN), Japanese Standards Association (JSA), and Battery Safety Organization (BATSO). It is beyond the scope of this book to discuss all potential standards and protocols; however, a summary of the many standards and testing protocols for Li-ion cells has been published previously.[15]

2.4 ESS Codes and Standards

Exponent reviewed relevant codes and standards relating to the design, testing, and installation of Li-ion ESSs.

2.4.1 Safety Standards

In addition to the numerous standards and protocols available for Li-ion batteries, there are a number of safety standards for the overall construction of Li-ion stationary battery systems and ESSs. These safety standards generally include a minimum set of construction requirements with which the system should comply, as well as a number of performance tests to ensure the system will operate safely over its anticipated life. These construction requirements typically address some or all of the following: material choices/requirements; electrical spacing of components; wiring criteria; controls and other components; failure modes and effects analysis (FMEA); and functional safety requirements, markings, signage, and instructions.

Performance tests are conducted to ensure that the Li-ion battery ESS operates safely under normal use and foreseeable misuse conditions. Some examples of performance tests include: normal operation at a variety of expected temperatures; anticipated abnormal events, such as short circuit tests or other tests for foreseeable fault conditions; electrical spacing and insulation tests, such as a dielectric voltage

[15]UL, "Safety Issues for Lithium-Ion Batteries," 2012.

test; and environmental conditions, such as exposure to water or other environmental stresses.

The published safety standards for Li-ion ESSs are often divided into technology specific and/or application specific documents. Some standards are intended for specific countries or geographical regions, while others are written as international standards. For battery ESSs, many of these standards were written for more traditional technologies, such as lead acid or nickel-cadmium (NiCad) battery systems and many of the documents are in the form of guides or recommended practices rather than standards; however, they still contain valuable information for evaluating and determining the safety of the ESS. It is beyond the scope of this book to discuss in detail all of the potential standards, guides, and recommended practices; however, a summary of many testing protocols for stationary battery systems and ESSs has been published previously.[16] The following is a list of many of the relevant documents and a brief summary of those documents that directly apply[17] to Li-ion battery ESSs and/or stationary battery systems:

- UL 1973, *Batteries for Use in Light Electric Rail (LER) and Stationary Applications* (UL 1973), is a safety standard for stationary batteries for energy storage applications that is not specific to any one battery technology or chemistry, and can apply to Li-ion battery ESSs, as well as ESSs using other battery chemistries. The standard includes construction requirements, safety performance tests, and production tests.[18] The Li-ion batteries assessed in the testing described in this book are listed to UL 1973.
  UL 1973 contains a series of construction parameters, including requirements for non-metallic materials, metallic parts resisting corrosion, enclosures, wiring and terminals, electrical spacing and separation of circuits, insulation and protective grounding, protective circuits and controls, cooling/thermal management, electrolyte containment, battery cell construction, and system safety analyses.
  UL 1973 also outlines a series of safety performance tests for ESSs, including electrical tests such as an overcharge test, short circuit test, over-discharge protection test, temperature and operating limits check test, imbalanced charging test, dielectric voltage test, continuity test, failure of cooling/thermal stability system test, and working voltage measurements. In addition, UL 1973 requires testing of electrical components, including a locked-rotor test for low voltage

[16]UL, "Draft Storage/Stationary Batteries Standards List.".

[17]Other documents that apply to battery ESSs or stationary battery systems that do not include Li-ion technologies within their scope were reviewed. Examples of such documents include: IEC 62485-2, *Safety Requirements for Secondary batteries and battery installations: Part 2 stationary*; IEC 60896-11, *Stationary lead-acid batteries Part 11: Vented types—General requirements and methods of tests*; IEC 60896-22, *Stationary lead-acid batteries Part 22: Valve regulated types—Requirements*; IEC 60896-21, *Stationary lead-acid batteries Part 21: Valve regulated types—Methods of test*; EN50272-2, *Safety Requirements for Secondary batteries and battery installations: Part 2 stationary*.

[18]UL 1973, *Batteries for Use in Light Electric Rail (LER) and Stationary Applications*.

direct current (DC) fans/motors in secondary circuits, input, leakage current, a strain relief test and a push-back relief test.

Mechanical tests are also required by UL 1973, including a vibration test, shock test, and crush test, which only apply to LER applications. Other mechanical tests that apply to all systems include a static force test, impact test, drop impact test, wall mount fixture/handle test, mold stress test, pressure release test, and a start-to-discharge test.

Additional environmental tests are also required by UL 1973, including a thermal cycling test, resistance to moisture test, and a salt fog test.

Of particular relevance to this study, UL 1973 also requires two fire exposure tests: an external fire exposure test and an internal fire exposure test. The purpose of the external fire test is to ensure that an ESS will not explode as a result of being exposed to a hydrocarbon pool/brush fire. In the external test, a fully charged ESS is subjected to a heptane pool fire, or another similar hydrocarbon fuel pool fire, for 20 min. The fuel is held in a pan placed 24 in. under the ESS and is sized (in diameter) to be large enough to cover the dimensions of the ESS. After the 20 min exposure, the ESS is subjected to a hose down in accordance with UL 263, *Conduct of Hose Stream Test of the Standard for Fire Tests of Building Construction and Materials*, to represent the firefighter response that the system may be exposed to during a fire. The ESS must demonstrate that no explosion hazards exist by the observation and measurement of any projectiles that occur during the external fire test.

The internal fire test is meant to demonstrate how the ESS will prevent a single cell failure within the battery system from cascading into a fire and/or explosion. In the internal fire test, the fully charged ESS is subjected to heating until thermal runaway of one internal battery cell that is centrally located within the ESS. Once the thermal runaway is initiated, the mechanism used to create thermal runaway is shut off or stopped and the ESS is subjected to a 1 h observation period. Fire cannot propagate during this observation period or result in an explosion.

- IEC 61,427-1, *Secondary cells and batteries for renewable energy storage—General requirements and methods of test—Part 1: Photovoltaic off-grid applications*, provides general information relating to the requirements for the secondary batteries used in photovoltaic energy systems (PVES) and the typical test methods used for the verification of battery performance. This standard deals with cells and batteries used in photovoltaic off-grid applications and is applicable to all types of secondary batteries, including Li-ion.[19]
- IEC 61427-2, *Secondary cells and batteries for renewable energy storage—General requirements and methods of test—Part 2: On-grid applications*, is a standard currently under development by IEC that relates to secondary batteries used in on-grid electrical energy storage (EES) applications. It provides test methods for the verification of their endurance, properties, and electrical

---

[19]IEC 61427-1, *Secondary cells and batteries for renewable energy storage—General requirements and methods of test—Part 1: Photovoltaic off-grid applications*, 2013 Edition.

performance in such applications. The test methods are essentially battery chemistry neutral, i.e., applicable to all secondary battery types, including Li-ion. On-grid applications are characterized by the fact that batteries are connected via power conversion devices to a regional, nation-, or continent-wide electricity grid and act as instantaneous energy sources and sinks to stabilize the grid's performance when major amounts of electrical energy from renewable energy sources are fed into it.[20]

- IEC 62619, *Secondary cells and batteries containing alkaline or other non-acid electrolytes—Safety requirements for large format secondary lithium cells and batteries for use in industrial applications*, is under development by IEC and will provide requirements on safety aspects associated with the erection, use, inspection, maintenance and disposal of cells and batteries for stationary applications and motive (other than on-road vehicles). It includes safety requirements for Li-ion cells for stationary and off-road motive applications and some battery requirements (evaluation of battery and battery management system [BMS] combination). The standard is not a system standard however, as it covers only battery and BMS interactions.

Two standards are currently under development by UL and the IEC that, when finished, will directly apply to commercial and residential Li-ion battery ESSs, including:

- UL Subject 9540, *Outline of Investigation for Energy Storage Systems and Equipment* (UL 9540), which will cover various types of ESSs and is not specific to just one battery chemistry or technology. Its scope includes requirements for ESSs that are intended to store energy from power or other sources and provide electrical or other types of energy to loads or power conversion equipment. The ESSs may include equipment for charging, discharging, control, protection, communication, controlling the system environment, fuel or other fluid movement and containment. The system may be standalone to provide energy for local loads or can be in parallel with an electric power system, electric utility grid or applications that perform multiple operational modes. The standard contains a series of construction parameters with material flammability criteria and performance tests for ESSs. Although no full-scale fire test of the ESS as an assembly is required, UL 9540 does require that Li-ion ESSs meet the requirements of UL 1973, which contains two fire tests, as described previously.[21]
- IEC 62897, *Stationary Energy Storage Systems with Lithium Batteries—Safety Requirements*, is under development by IEC and will provide general safety requirements for stationary ESSs with lithium batteries. The standard will

[20]IEC 61427-2, *Secondary cells and batteries for renewable energy storage—General requirements and methods of test—Part 2: On-grid applications*, 2015 Edition.

[21]UL 9540, *Outline of Investigation for Energy Storage Systems and Equipment*, Issue Number 1, June 30, 2014.

incorporate a number of requirements to address potential hazards with ESSs, including: electric shock or burn; mechanical hazards; spread of fire from the equipment; excessive temperature; effects of fluids and fluid pressure; liberated gases, explosion; and chemical hazards (e.g., electrolyte). The standard intends to cover small battery systems for residential or similar use that can be connected to a main source of supply.[22, 23]

2.4.2 *Codes and Regulations*

In addition to safety standards, there are local, state, and national electrical, building, and fire codes to consider that could impact the installation of ESSs. In the United States, the codes affecting ESSs include the electrical installation codes, such as NFPA 70, *National Electrical Code* (NEC) and fire codes, such as NFPA 1, *Fire Code* (NFPA 1) or the International Code Council (ICC) code suite for building and fire codes. Electrical codes, such as the NEC, include requirements, among others, for wiring methods, grounding criteria, signage, and enclosures that impact ESS electrical safety.[24] Building and fire codes include requirements for battery rooms, spill containment, and fire protection systems for areas containing battery storage that impact the fire risk of the building, its occupants, and contents.

Concerns have arisen from the perceived lack of information contained in local, state, and national codes and regulations as they relate to Li-ion ESSs. Some of the concerns include: (1) limited information in the codes specifically relating to Li-ion batteries; (2) volume of electrolyte in the Li-ion battery being used to define its hazard level (which is not appropriate for Li-ion battery chemistry[25]); (3) fire suppression and detection systems required to protect ESSs; (4) whether or not these batteries are considered hazardous materials; and (5) separation of ESSs from other portions of the building.

2.4.2.1 Electrical Codes

NEC Article 480, *Storage Batteries*, applies to all stationary installations of storage batteries. Article 480 was originally written for and generally applied to stationary

---

[22]http://www.iec.ch/dyn/www/f?p=103:38:0::::FSP_ORG_ID,FSP_APEX_PAGE,FSP_LANG_ID,FSP_PROJECT:1410,23,25,IEC%2062897%20Ed.%201.0.

[23]UL, "Draft Storage/Stationary Batteries Standards List."

[24]NFPA 70, 2014 Edition, Article 480, *Storage Batteries.*

[25]There are a number of reasons why the "volume of electrolyte" is not appropriate. One example is that the volume of electrolyte inside a battery cell is not extractable from a completed cell; therefore, the volume of electrolyte inside a Li-ion cell does not meaningfully translate to a hazard. The volume of electrolyte is appropriate for other chemistries, such as lead acid, where the failure of a battery could lead to spilling of the aqueous solution; however, the failure of a Li-ion battery or cell will more likely lead to the venting of a flammable gas, not the release of a liquid.

lead acid battery installations in the range of 48 V. The section outlines a series of requirements for battery installations, however, most pertain to the electrical safety of the systems and have limited requirements specific to fire protection that would address the industry concerns listed above. For example, the NEC has sections on battery and cell terminations (Section 480.3), wiring and equipment supplied from batteries (Section 480.4), overcurrent protection (Section 480.5), disconnect methods (Section 480.6), insulation (Section 480.7), racks and trays that support the batteries (Section 480.8), battery locations (Section 480.9 Parts (A), (B), and (G)), and safety vents[26] (Section 480.10). Section 480.9, *Battery Locations,* Parts (C) and (D) requires certain working spaces clearances for battery systems to allow for the units to be properly accessed. In addition, Part (E) requires that personnel door(s) intended for entrance to and egress from rooms designated as battery rooms open in the direction of egress and be equipped with listed panic hardware. Gas piping is also prohibited from being installed within a dedicated battery room in Section 480.9 Part (F).

The next edition of the NEC to be published, the 2017 edition, is proposed to have a new article (Article 706) dedicated to ESSs. This addition should further assist installers, AHJs and manufacturers with navigating the electrical installation requirements for these systems.

2.4.2.2 Building and Fire Codes

Below is a summary of the sections contained within the 2015 edition of the International Building Code (IBC), International Residential Code (IRC), International Fire Code (IFC), and NFPA 1 relating to Li-ion ESSs and the concerns listed above. Many of the identified gaps in the codes mentioned below are currently being worked on and may be addressed when the next round of codes are published.

1. **Limited information on Li-ion battery ESSs**. Recent additions to the building and fire codes have answered many industry concerns, providing more details and thresholds for when requirements are necessary for Li-ion battery systems. Starting in 2006 for the IFC and 2009 for NFPA 1,[27, 28] Li-ion batteries for use in stationary storage battery systems were discussed. Many municipalities lag behind in the adoption of new editions of building and fire codes. As such, those areas still using older versions of the codes could encounter issues; however, this issue (besides the correlating issues highlighted below in #2) is one that should resolve itself with the adoption of the newer codes.

[26]Li-ion batteries do not typically require venting due to their technology and design, which does not vent hydrogen.

[27]2006 IFC, Section 608.1.

[28]NFPA 1, 2009 Edition, Section 52.1.

The 2015 edition of the IRC does not contain language relating to stationary battery systems, ESSs, or other similar systems, which could be confusing for readers looking for guidance for systems being installed in one or two-family dwellings or townhouses.

2. **Volume of electrolyte**. Traditionally, the IBC, IFC, and NFPA 1 applied specific safety requirements to battery systems containing more than 50 gallons of electrolyte. However, this requirement cannot be applied to Li-ion battery systems, as the electrolyte is not stored in an aqueous solution. To account for this, starting in 2006 for the IFC and 2009 for NFPA 1, the fire codes defined the threshold at which requirements are necessary for Li-ion stationary storage battery systems according to their weight (1000 lb).[29, 30] Adding to some of the confusion in the marketplace when discussing Li-ion battery packs and how best to define/categorize them, other agencies beyond the ICC and NFPA also utilize varying methods. For instance, the United Nations, *Recommendations on the Transport of Dangerous Goods—Manual of Tests and Criteria*, also defines and categorizes batteries by mass, where anything larger than 12 kilograms (kg) of gross mass is a "large battery" and anything less than 12 kg is a "small battery." In addition, a "large cell" is defined as anything with a gross mass greater than 500 grams (g). A cell less than 500 g is considered a "small cell."[31] The Department of Transportation (DOT) in 49 CFR 173.185 defines and categorizes batteries by "equivalent lithium content" (ELC), where the ELC is the product of the rated capacity, in Ah, of a Li-ion cell times 0.3, with the result expressed in grams. The ELC for a battery pack equals the sum of the grams of ELC contained in the component cells of the battery.[32] As such, DOT categorizes Li-ion batteries by their capacity, not the volume of electrolyte or mass of the cell or battery pack.

   Even with the addition of the weight threshold for Li-ion battery systems in 2006 and 2009, the IBC, IFC, and NFPA 1 each still contain language in other sections of the codes that discuss requirements when the volume of electrolyte is above the 50-gallon threshold, not taking into account the weight of a Li-ion battery system. Three instances identified in the codes where this occurs include:

   (a) IBC Section 907.2.23, which states that any battery room with greater than 50 gallons of electrolyte must have a smoke detection system. IFC Section 608 applies directly to stationary storage battery systems and Li-ion batteries and resolves any confusion that exists in the code, as Section 608.9 requires a smoke detection system for stationary battery systems that are

[29]2006 IFC, Section 608.1.

[30]NFPA 1, 2009 Edition, Section 52.1.

[31]United Nations, *Recommendations on the Transport of Dangerous Goods—Manual of Tests and Criteria.*

[32]49 CFR 171.8.

large enough to trigger the thresholds, such as a Li-ion battery system greater than 1000 lb.[33] However, if a reader were to miss that section of the IFC, and only read the section in the IBC, it could create confusion over how to apply Section 907.2.23 to Li-ion battery systems.

(b) IFC Section 105.7.2, which states that battery systems with more than 50 gallons of electrolyte require a permit before installation. However, no weight threshold is provided for Li-ion batteries.[34] As such, there could be confusion regarding whether or not a permit is required for Li-ion battery systems.

(c) NFPA 1 Table 1.12.8(a), which states that lead-acid battery systems with more than 50 gallons (unsprinklered buildings) or 100 gallons (spinklered buildings) of electrolyte require a permit before installation. However, Li-ion battery systems are not addressed in Table 1.12.8(a).[35] As such, there could be confusion regarding whether or not a permit is required for Li-ion battery systems.

3. **Suppression and detection**. Where required, such as for a high-rise building, fire sprinklers are not required in the area where battery systems are installed, provided the space is equipped with an automatic fire detection system and is separated from the rest of the building with 1 h barriers or 2 h horizontal assemblies.[36] In addition, a smoke detection system is required for all Li-ion battery systems greater than 1000 lb.[37, 38]
4. **Hazardous materials**. The IBC and NFPA 1 state that battery systems do not fall into the Hazardous Group H category (for the IBC) or should be considered a hazardous material (for NFPA 1) provided certain ventilation requirements for the ESS are met.[39, 40] However, Li-ion batteries typically do not require room ventilation,[41] as off gassing does not occur during normal operation. It is unclear if the IBC and NFPA 1 requirement for room ventilation is necessary for a Li-ion battery ESS to ensure it does not fall into the hazardous category.
5. **Separation**. The IBC states that Li-ion battery systems more than 1000 lb in weight shall be separated from the remainder of the building by either a 1 h separation or 2 h separation depending on the occupancy in which it is installed.[42]

---

[33]2015 IFC, Section 608.1 and 608.9.

[34]2015 IFC, Section 105.7.2.

[35]NFPA 1, 2015 Edition, Table 1.12.8(a).

[36]2015 IBC, Section 403.3 and Exception to Section 903.2.

[37]2015 IFC, Section 608.9.

[38]NFPA 1, 2015 Edition, Section 52.3.10.

[39]2015 IBC, Section 307.1.1(9).

[40]NFPA 1, 2015 Edition, Section 60.1.2.

[41]2015 IFC, Table 608.1 and NFPA 1, 2015 Edition, Table 52.1.

[42]2015 IBC, Table 509.

Section 608 of the IFC and Chapter 52 of NFPA 1 provide further guidance on the proper installation of Li-ion ESSs. However, many of the requirements do not apply to Li-ion due to the chemistry of battery, including safety caps, spill control and neutralization measures, and room ventilation. Signage, seismic protection, and a fire/smoke detection system are required for Li-ion battery systems larger than 1000 lb.[43, 44] A review of these two sections also identified another potential area of confusion for a user of the codes. The IFC does not require thermal runway protection for Li-ion battery systems, while NFPA contains contradictory guidance. Thermal runaway can occur in Li-ion battery systems and it is unclear why thermal runaway protection in Li-ion battery systems is not required in the IFC. NFPA 1 Table 52.1 states that Li-ion battery systems do not require thermal runaway protection; however, Sect. 52.3.2 states that Li-ion battery systems, "shall be provided with a listed device or other approved method to preclude, detect, and control thermal runaway." Table 52.1 and the language of Sect. 52.3.2 are in direct conflict with one another, leading to possible confusion for anyone using the code. A review of the Report on Proposals and Report on Comments from the 2009 NFPA 1 code development cycle provided some guidance regarding what the technical committee intended. It appears that the technical committee intended for the thermal runaway protection to be required; however, a typo in Table 52.1 was not fixed at the time of initial adoption or anytime during future code development cycles. This issue should be addressed in the next code development cycle to remove any confusion as to what NFPA 1 requires for thermal runaway protection of Li-ion battery systems.

2.5 ESS Fire Incidents

A review of fire incidents reported in the National Fire Incident Reporting System (NFIRS) from 1999 to 2013 was performed during the literature review. During this time period, only 44 % of fires that fire departments respond to were captured in NFIRS. Thus, the numbers listed below do not account for every fire in the United States during that time. In addition, NFIRS currently does not have a means to report a stationary battery system or ESS fire; however, they do have a coding system for uninterrupted power supplies (UPS; code 226) and batteries (code 229). Table 2.1 provides a summary of the number of UPS and battery fires that were reported in NFIRS between 1999 and 2013.

Exponent also searched for public incidents tied directly to the involvement of Li-ion ESSs in a fire. Through this search, only two major events involving battery ESSs were identified, one at a wind turbine power generating facility in Hawaii and one at a solar energy facility in Arizona. However, only the Arizona facility contained a Li-ion battery ESS, which was a pilot ESS that the facility was testing.

[43]2015 IFC, Table 608.1.

[44]NFPA 1, 2015 Edition, Table 52.1.

Table 2.1 Summary of NFIRS data

Incident type	UPS fire (Code 226)	Battery fire (Code 229)
Structure fire or fire in mobile property used as a fixed structure	142	318
All fires (not just structures)	227	1014

Summaries of these two incidents ascertained from public sources are provided in the following sections.

No publically reported fire incidents were identified to have started in or significantly involved a commercial or residential ESS

2.5.1 *Kahuku Wind Energy Storage Farm Battery ESS Fires*

Three fires occurred at the Kahuku Wind Energy Storage Farm over the course of a year and a half span from April 2011 to August 2012. The ESS contained 12,000 individual lead acid battery packs for a capacity of 15 megawatts (MW). The battery packs were stacked 6 ft high inside a 9000 ft^2 metal warehouse building. It was determined that the fires were caused by undersized capacitors used by the battery system. The first two fires were allowed to self-extinguish, with limited damage to the system and the building; however, the third fire resulted in a total loss of the building and contents, including the 12,000 battery packs.[45, 46, 47]

The first incident occurred on April 22, 2011; the alarm was received by dispatch at approximately 5:45 p.m. and the Honolulu Fire Department (HFD) arrived on scene approximately 10 min later. An engineer from Xtreme Power, Inc. (Xtreme) was alerted by a remote alarm indicating that an exhaust fan on the Hawaiian Electric Company (HECO) side of the structure had overheated. The engineer also stated that smoke and popping sounds were emanating from the structure before HFD arrived. When HFD arrived, they noted smoke coming from the battery storage building. Approximately an hour after the first alarm, other arriving HFD personnel reported moderate grayish black smoke emanating from the structure, with no flames visible and no other structures in immediate danger. Facility personnel provided battery safety data sheets (SDS) for the lead acid batteries and building plans, however, HFD chose to wait for daylight to make an interior attack,

[45]http://www.windpowermonthly.com/article/1284038/analysis-first-wind-project-avoids-storage-30m-fire.

[46]http://www.greentechmedia.com/articles/read/Battery-Room-Fire-at-Kahuku-Wind-Energy-Storage-Farm.

[47]http://www.scientificamerican.com/article/battery-fires-pose-new-risks-to-firefighters/.

primarily due to concerns regarding the stored energy in the batteries and possibly unsafe night operations. Major hazards identified by Xtreme and HFD included the batteries themselves (possibly explosive or energized), the sulfuric acid from the batteries, toxic environment, and energized electrical equipment. Xtreme advised HFD that water could not be used to extinguish the fire and that dry chemical, carbon dioxide (CO_2), or specialty foam (FM200) would be the best extinguishing agent. HECO personnel arrived on scene to secure the power to the building and advised of a sulfuric acid odor at the HECO switch box, emanating from the conduits within the building. A firewatch was present throughout the night. The following day, HFD made entry into the building, but no active burning was found. The building was ventilated and cleared and operators of the facility were allowed to investigate and notified HFD that the cause of the incident was a failed electrical inverter. HFD investigators concluded that the origin of the fire was in the battery ESS building, within the Inverter #9 cabinet. The first material ignited was most likely conductor insulation or associated components within the cabinet. The fire was classified as accidental, failure and/or malfunction of operating electrical equipment. Fire spread was confined to the object of origin.[48]

The second incident occurred on May 23, 2011; the alarm was received by dispatch at approximately 10:20 p.m. and HFD arrived on scene approximately 10 min later. When HFD arrived, they noted light smoke coming from the top of a roll up door at the same ESS building. Facility personnel advised HFD that the incident appeared to be the same as the first loss; therefore, the same actions were taken, including shutting down the power and closing the building until morning. The next morning, HFD arrived to no smoke. The building was ventilated and one inverter was found to be burned out, with no residual signs of heat.[49]

The third incident occurred on August 1, 2012; the alarm was received by dispatch at 4:44 a.m. and HFD arrived on scene approximately 15 min later. First Wind advised HFD that their sensors indicated the malfunction of an electrical inverter directly adjacent to the stacks of batteries in the ESS building. Due to the large amount of batteries stored on site and experiences in the prior incidents, HFD chose to standby and monitor the building until HECO arrived with their dry chemical extinguishing truck. The fire was monitored using a thermal imaging camera and smoke and heat intensified, eventually venting through the roof, with some flames visible. Water was used to cool the uninvolved side of the building, but was discontinued due to the risk of contact with the burning batteries. Once HECO arrived, HFD assisted with deploying the dry chemical extinguishing line; however, suppression efforts were unsuccessful, as the dry chemical could not reach all of the burning material and entry could not be made due to the hazardous conditions created by the burning batteries and lack of an adequate supply of dry chemical. The fire eventually involved the entire building. Water was used to prevent spread to adjacent buildings, however, water could not be applied to the incident building

[48]Honolulu Fire Department Incident Report 2011-0018972.

[49]Honolulu Fire Department Incident Report 2011-0023875.

due to environmental concerns regarding runoff water, as well as the high potential for stored electrical energy in the malfunctioning system and the large quantities of sulfuric acid involved; therefore, the fire was contained to the original building and allowed to burn until it eventually self-extinguished. HFD noted that significant/unusual fuel load from contents was a factor in suppressing the fire; with the material contributing most to flame spread being plastic used as electrical wire, cable insulation. HFD investigators concluded that the origin of the fire was in the battery ESS building. The first alarm activation was within the Inverter #9 cabinet, followed by general building smoke alarm activation. Video taken inside the ESS building showed fire in the proximity of the Inverter #9 cabinet. The first material ignited was most likely conductor insulation or associated components within the cabinet. The physical construction of the 12,000 batteries and associated conductors contributed mostly to fire spread. The fire was classified as accidental, failure and/or malfunction of operating electrical equipment.[50]

These fires demonstrate the need for better understanding of ESS fires so that the owner and fire departments responding to these incidents can better prepared in the event of a fire.

2.5.2 *Arizona Public Service Company ESS Fire*

In November of 2012, a fire occurred at a state-of-the-art solar energy storage system the Arizona Public Service Company (APS) was testing. The system, the relative size of a shipping container with a capacity of 1.5 MW, had been running since February of 2012. Similar to the First Wind fires, fire department personnel allowed the fire to burn freely for some time. The cause of the fire was not reported.[51, 52] Exponent requested the local fire department reports on these fire incidents to obtain further details of the incidents, however, no response was received.

To date, relatively few ESS systems have been commissioned. In addition, most systems commissioned have been lead acid battery systems, not Li-ion. The search for fires involving ESSs has identified only a few from publically available sources. In order to gain insight into how Li-ion ESSs will behave in fire scenarios, we can examine fires involving similar systems or battery fires in general.

[50]Honolulu Fire Department Incident Report 2012-0038895.

[51]http://www.energy-storage-online.com/cipp/md_energy/custom/pub/content,oid,1133/lang,2/ticket,g_u_e_s_t/~/APS_fire_probed.html.

[52]http://azdailysun.com/news/local/aps-fire-probed/article_1de2e924-ab0a-5e71-9a3a-6942c2d1c9bb.html.

2.6 Li-ion Battery Fires

Given the lack of ESS fire incidents documented in the literature, a review of Li-ion battery fires was conducted. Fires may occur in an ESS high voltage battery, or a fire may extend to the battery, attacking the ESS from the outside in. Previous research programs have been conducted focusing on large format Li-ion battery fires, electric drive vehicle (EDV) Li-ion battery fires, and Li-ion battery storage fires. This research involved full-scale fire tests of Li-ion batteries that were polymer, prismatic, and cylindrical designs.

For large format Li-ion battery systems with polymer or prismatic designs, the research has generally shown the following hazards associated with fires:

1. Fire tests of identical vehicles indicated that the heat release rate (HRR) of an EDV compared to a more common internal combustion engine (ICE) vehicle are similar [8] and a free burn (no suppression) test of an EDV battery did not produce significant HRRs [4].
2. Test results indicate that water can be an effective extinguishing agent on large format Li-ion battery fires, however, large quantities may be required for extinguishment [4, 9].[53]
3. During fires tests of EDVs with polymer pouch battery cells, no projectiles or explosions from the large format batteries were observed [4, 8, 10].
4. Gas samples collected during fire tests of complete (i.e., full) ICE vehicles and EDVs identified similar levels of toxic compounds in the smoke, including CO_2, nitrogen oxides (NO_x), hydrogen cyanide (HCN), hydrogen chloride (HCl), carbon monoxide (CO), and hydrogen fluoride (HF) [8]. In addition, water samples collected after extinguishing Li-ion batteries showed concentrations of fluoride and chloride [4, 9].
5. Fire tests have also demonstrated that in the tested scenario, with a battery pack tested inside a vehicle fire trainer (i.e., not a powered consumer EDV), the shock/electrocution hazards of applying a water stream directly to an energized high voltage battery that has been compromised by heat and fire were negligible [4]. In addition, other fire tests where hose streams were applied directly to energized electrical equipment have demonstrated that current leakage through the suppression water is not a hazard, provided sufficient clearance distances for the given voltage of the electrical equipment are observed between the hose stream and conductors [11–13].[54]

Previous research focusing on large format Li-ion battery fires with a polymer or prismatic design demonstrated that some of the common concerns regarding Li-ion battery fires (namely explosions, projectiles, and toxic gas formation) have not been replicated in full-scale fire tests. However, fire tests of unconfined Li-ion batteries

[53]Delphi Corporation. Hybrid Electric Vehicles for First Responders. Troy, MI. 2012.

[54]Factory Mutual Handbook of Industrial Loss Prevention, "Electrical Conductivity of Extinguishing Agents".

with a cylindrical design have demonstrated that "cell explosions" can occur with projectiles observed traveling up to 133 ft [14].

2.7 Gap Analysis

Based upon the literature review conducted to date, Exponent has identified the following gaps in the knowledge base for commercial and residential Li-ion ESSs:

1. No public fire test data demonstrating the fire behavior of ESSs.
2. Limited public fire test data related to large format battery packs with cylindrical design utilized either in vehicles or storage systems.
3. No fire test data or publically available real world fire incidents involving residential or commercial Li-ion ESSs illustrating the hazards (projectiles, heat release, toxic gas production) to first responders and/or the best practices for fire department operations.
4. Limited real world fire incidents involving large-scale (grid size) ESSs.
5. No Li-ion ESS guidance in the IRC.
6. Some sections of the IBC, IFC, and NFPA 1 are confusing, as only the volume of the electrolyte (a requirement for older battery chemistries such as lead acid) and not the weight of the Li-ion battery system, is used as a threshold for when certain building or fire code requirements are necessary. In addition, other agencies, such as the United Nations and DOT, have other methods for defining and categorizing batteries. Many of these code sections are presently being revised and could be addressed by the next published code set.
7. NFPA 1 provides contradictory guidance regarding thermal runway protection for Li-ion battery systems, while the IFC does not require thermal runaway protection for Li-ion battery systems at all. Many of these code sections are presently being addressed and could be resolved by the next published code set.
8. No post-fire incident response and recovery (i.e., overhaul) procedures.
9. No stationary battery system or ESS fire reporting code in NFIRS to assist in analyzing fire incidents and differentiate battery systems from household batteries.

References

1. Wald, Matthew, L. Wind Drives Growing Use of Batteries, *The New York Times*, July 27, 2010.
2. Mikolajczak, C., Kahn, M., White, K., and Long, RT. "Lithium-Ion Batteries Hazard and Use Assessment." Fire Protection Research Foundation Report, July 2011.
3. Long RT and Mikolajczak CJ. "Lithium-ion batteries hazards: What you need to know." Fire Protection Engineering Q4 2012.

4. Long RT, Blum AF, Bress TJ, and Cotts BRT. "Emergency response to incidents involving electric vehicle battery hazards." Fire Protection Research Foundation Report, July 2013.
5. Long RT, Sutula JA, and Kahn MJ. "Lithium-ion batteries hazard and use assessment Phase IIb." Fire Protection Research Foundation Report, 2013.
6. Dahn J, Ehrlich GM, "Lithium-Ion Batteries," *Linden's Handbook of Batteries*, 4th Edition, TB Reddy (ed), McGraw Hill, NY, 2011.
7. Balakrishnan PG, Ramesh R, Prem Kumar T, "Safety mechanisms in lithium-ion batteries," Journal of Power Source, 155 (2006), 401-414.
8. Lecocq, A, Bertana M, Truchot, B. and Marlair G. "Comparison of the Fire Consequences of an Electric Vehicle and an Internal Combustion Engine Vehicle." INERIS – National Institute of Industrial Environment and Risks, Verneuil-en-Halatte, France. Second International Conference on Fires in Vehicles, September 27-28, 2012, Chicago, IL.
9. Egelhaaf, M., Kress, D., Wolpert, D., Lange, T., Justen, R., and Wilstermann, H., "Fire Fighting of Li-Ion Traction Batteries," SAE Int. J. Alt. Power. 2(1):37-48, 2013, doi: 10.4271/2013-01-0213.
10. Watanabe, N., Sugawa, O., Suwa, T., Ogawa, Y., Hiramatsua, M., Tomonoria, H., Miyamotoa, H., Okamotoa, K., and Honmaa, M. "Comparison of fire behaviors of an electric-battery-powered vehicle and gasoline-powered vehicle in a real-scale fire test." National Research Institute of Police Science, Japan. Presented at Second International Conference on Fires in Vehicles, September 27-28, 2012, Chicago, IL.
11. Sprague, C.S. and C.F. Harding. "Electrical Conductivity of Fire Streams" Research series no. 53. Engineering Experiment Station, Purdue University Lafayette, Indiana, January 1936.
12. Bolander, G.G., Jughes, J. T., Toomey, T. A., Carhart, H.W., and J.T. Leonard. "Use of Seawater for Fighting Electrical Fires" Navy Technology Center for Safety and Survivability, Chemistry Division. May 25, 1989.
13. Backstrom, R., Dini, DA, "Firefighter Safety and Photovoltaic Installations Research Project." Underwriters Laboratories Inc. November, 2011.
14. Webster, H, "Preliminary Full-Scale Fire Tests with Bulk Shipments of Lithium Batteries." 2012 FAA Fire Safety Highlights, US Department of Transportation Federal Aviation Administration, 2012.

Chapter 3
Testing Program Summary

Exponent, in conjunction with the Project Technical Panel, their advisory groups, and industry sources, identified and procured two (2) Li-ion battery ESSs for full-scale testing. The battery pack utilized in the ESS is a 100 kWh unit manufactured by Tesla Energy (Tesla) meant for commercial applications (Powerpack). The Powerpack consists of a 52-in. long by 38-in. wide by 86-in. tall steel cabinet containing the battery, protection electronics, and thermal management systems. The total weight of the unit is 3970 lb and it mounts directly to a concrete pad. A more detailed description of the ESS tested is provided in Chap. 4.

The full-scale fire tests were separated into two categories: (1) external ignition of the Powerpack and (2) internal ignition of the Powerpack. During the external ignition test, the Powerpack was exposed to an external fire source (a propane burner) to simulate a fire scenario where a fire originates outside of the Powerpack. During the internal ignition test, individual battery cells within the Powerpack were forced into thermal runaway.

3.1 Test Instrumentation Summary

Both tests were performed outdoors in open air, on a concrete pad, exposed to natural weather conditions, as would be typical of an outdoor commercial installation. In the external ignition testing, a propane burner system was used to apply the thermal assault to the Powerpack and cause thermal runaway of the batteries within. During internal ignition testing, the batteries of the Powerpack were forced into thermal runaway at the individual cell level.

Data collected during the tests included:

- Internal and external Powerpack surface temperatures;
- Heat fluxes at varying stand-off distances from the Powerpack;
- Internal Powerpack cabinet and pod pressures;

A.F. Blum and R.T Long Jr., *Fire Hazard Assessment of Lithium Ion Battery Energy Storage Systems*, SpringerBriefs in Fire,
DOI 10.1007/978-1-4939-6556-4_3

- Select products of combustion;
- Weather conditions;
- Projectile observations;
- Still photography; and
- High definition video.

3.2 Full-Scale Fire Protocols

Exponent and Tesla created two protocols for the full-scale fire tests: one for the external ignition test and one for the internal ignition test.

3.2.1 External Ignition Testing

The test protocol for the external ignition testing was as follows:

1. The Powerpack was positioned and the test equipment was set up as described in Sect. 3.1.
2. The following background data was collected as a steady-state baseline for 3 min:
   (a) Thermocouples;
   (b) Heat flux gauges; and
   (c) Gas sampling.
3. High definition video recordings were started simultaneously with data collection.
4. After the 3-min baseline was established, the propane burners were ignited to provide a 400 kW[1] exposure.
5. The 400 kW exposure was continued for approximately 60 min. Once at least twenty (20) cell thermal runaways were confirmed audibly, the burner was turned off.
6. Once the burner was shut off at the end of the approximate 60-min 400 kW exposure, the progression of the Powerpack fire in the free burn state was monitored thereafter.
7. Visual observations of importance were recorded, including when smoke was first observed, when cells went into thermal runaway, smoke production/color, projectiles, when flames were first observed, height and severity of flames, etc.
8. Still photographs were recorded throughout the test, as appropriate.
9. Data collection continued until all signs of combustion ceased.

[1]HRR from the propane burners was determined based upon the flow rate of propane recorded by a mass flow meter during testing times the heat of combustion of propane.

3.2.2 *Internal Ignition Testing*

The test protocol for the internal ignition tests was as follows:

1. The Powerpack was positioned and the test equipment was setup as described in Sect. 3.1.
2. The following background data was collected as a steady-state baseline for approximately 1.5 min:
    (a) Thermocouples and
    (b) Gas sampling.
3. High definition video recordings were started simultaneously with data collection.
4. After the 1.5-min baseline was established, multiple Powerpack cells were forced into thermal runaway through the use of heater cartridges by Tesla.
5. Visual observations of importance were recorded, including when smoke was first observed, when cells went into thermal runaway, smoke production/color, projectiles, when flames were first observed, height and severity of flames, etc.
6. Still photographs were recorded throughout the test, as appropriate.
7. Data collection continued until all signs of thermal runaway ceased.

Chapter 4
ESS Description

This chapter provides an overview of the Powerpack (a 100 kWh commercial ESS) utilized for this testing program. The Powerpack can be a single standalone unit, as shown in Fig. 4.1, or installed side by side with multiple Powerpacks if additional storage capacity is desired, as shown in Fig. 4.2.

4.1 ESS Battery Pack

The cells utilized within the Powerpack are 3.6 V, 2.4 A h cylindrical 18,650 cells. Two modules, each consisting of approximately 450 cells, are connected and enclosed inside a steel cover to form one energy storage pod, as shown in Fig. 4.3. As such, one energy storage pod contains a total of two modules, or approximately 900 battery cells. Sixteen (16) energy storage pods are contained within the Powerpack cabinet for a total of approximately 14,400 battery cells within the Powerpack.

4.1.1 ESS Design Layout

The Powerpack was designed for commercial installations. Within the Powerpack, Li-ion cells are contained within energy storage pods. The energy storage pods are housed inside a 52-in. long by 38-in. wide by 86-in. high steel cabinet. The total weight is 3970 lb. The front door of the Powerpack cabinet provides access to each of the 16 energy storage pods, as shown in Fig. 4.4, and contains equipment designed to thermally cool the pods. The liquid cooling system pumps a 50 % water/50 % ethylene glycol mixture to each of the 16 energy storage pods, as shown in Fig. 4.5. The coolant pumps, reservoirs, and associated fans and radiators are mounted and contained within the front door of the Powerpack. A refrigerant

A.F. Blum and R.T Long Jr., *Fire Hazard Assessment of Lithium Ion Battery Energy Storage Systems*, SpringerBriefs in Fire,
DOI 10.1007/978-1-4939-6556-4_4

Fig. 4.1 Single standalone Powerpack (100 kWh commercial ESS)

Fig. 4.2 Multiple Powerpacks installed side by side in an array

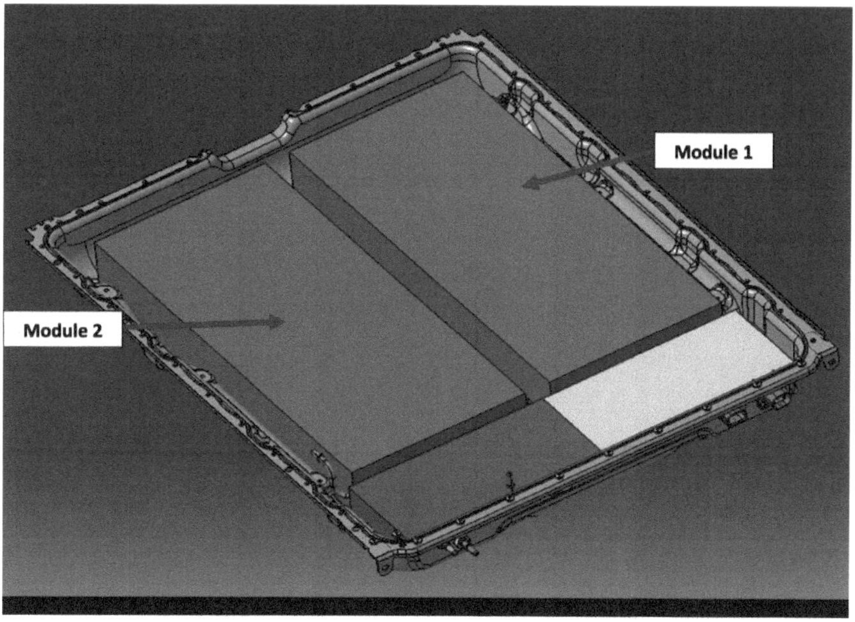

Fig. 4.3 Illustration of a pod with two modules (*blue*); each module contains approximately 450 cylindrical Li-ion battery cells totaling 900 battery cells for each pod (Color figure online)

system using 400 g of R134a further cools the ethylene glycol and is also mounted on the front door of the Powerpack. The back of the energy storage pods connect to an exhaust manifold at the rear of the Powerpack that has a vent at the top, as shown in Fig. 4.6.

4.2 ESS Safety Features

The Powerpack is listed to UL 1741, UL 1973, and IEC 62109. In addition, the Powerpack is designed to be compliant with UL 9540 and IEC 62619, currently under development by UL and IEC, respectively. UL1973, as described in Sect. 2.4.1, includes a number of construction requirements, performance tests, and production tests for stationary battery systems, including an external fire test and an internal fire test. The external fire test requires that the ESS not pose an explosion hazard if attacked by an external fire. The internal fire test demonstrates that a single battery cell failure within the center of the ESS battery pack will not result in a cascading thermal runaway of battery cells resulting in a propagating fire from the ESS and/or an explosion of the ESS.

Specific to the Powerpack design, each pod has a low voltage (approximately 50-V) output that is later converted through power management electronics into the

Fig. 4.4 Powerpack illustration (*left*) and image (*right*); depicting the sixteen (16) energy storage pods installed within the cabinet and instrumented for testing

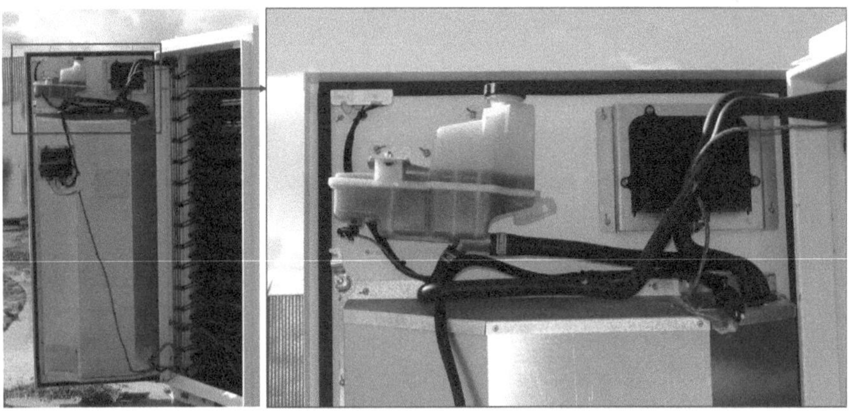

Fig. 4.5 Powerpack thermal door (*left*) and close up of the refrigerant (*right*)

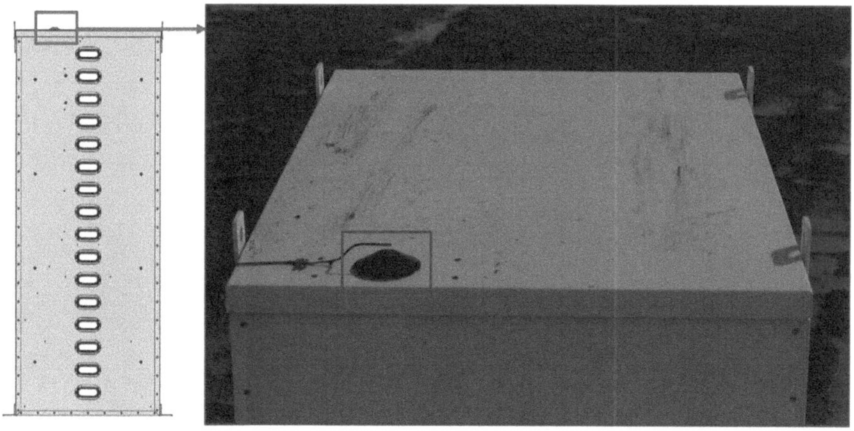

Fig. 4.6 Illustration of the Powerpack exhaust vent (*left*) and an image of the vent at the top of the Powerpack (*right*)

higher 400-V Powerpack output. The energy storage pods are galvanically isolated and the 400-V Powerpack output is only present when the Powerpack is in an active state and the power electronics are operational. Without active low voltage system electronics, because of the galvanic isolation, there is no electrical pathway from the live battery voltage to the exterior of a pod. As such, because of the design of the Powerpack, during charging or discharging, the cells are not at a high voltage. Each energy storage pod is encased inside a steel enclosure that prohibits any cell failure from projecting outside of pod. In addition, the pods are then enclosed within the steel Powerpack cabinet, which further reduces the possibility of projectiles from the unit. As described earlier, the energy storage pods are cooled by a thermal management system in the front door of the Powerpack cabinet that keeps the battery cells within safe operating temperatures. In the unlikely event of cell thermal runaway, the Powerpack has an engineered exhaust pathway, which directs runaway gas to a gas manifold that is directed out the top of the Powerpack. The Powerpack is designed to be installed side by side with multiple Powerpacks if additional storage capacity is desired. Clearance from the Powerpack is outlined in the manufacturer's installation manual, which requires that combustibles be kept six feet from the front, six inches from the sides and back, and five feet from the top of the Powerpack.

Chapter 5
Testing Setup

The full-scale fire tests were separated into two categories: (1) external ignition of the Powerpack and (2) internal ignition of the Powerpack, as described below. For both tests, the Powerpack battery packs were charged to a full 100 % state of charge (SOC) prior to testing.

5.1 External Ignition Testing

The external ignition test exposed the Powerpack to a propane burner to simulate a fire scenario where the fire originates outside of the Powerpack.

5.1.1 ESS Positioning

The Powerpack was positioned on a noncombustible surface similar to its intended end use for an outdoor installation on a concrete pad, as shown in Fig. 4.1. The test instrumentation, including thermocouples (TCs), heat flux gauges (HFGs), pressure transducers, gas sampling, data acquisition, weather meter, and cameras were positioned around the Powerpack as illustrated in Fig. 5.1. In addition, a propane burner, further described in Sect. 5.1.2, was placed to the right side of the Powerpack, allowing for direct flame impingement on the exterior of the Powerpack cabinet.

5.1.2 Burner Description (Ignition Source)

The burner assembly consisted of three (3) drilled-pipe propane burners connected in parallel, as shown in Fig. 5.2.

A.F. Blum and R.T Long Jr., *Fire Hazard Assessment of Lithium Ion Battery Energy Storage Systems*, SpringerBriefs in Fire,
DOI 10.1007/978-1-4939-6556-4_5

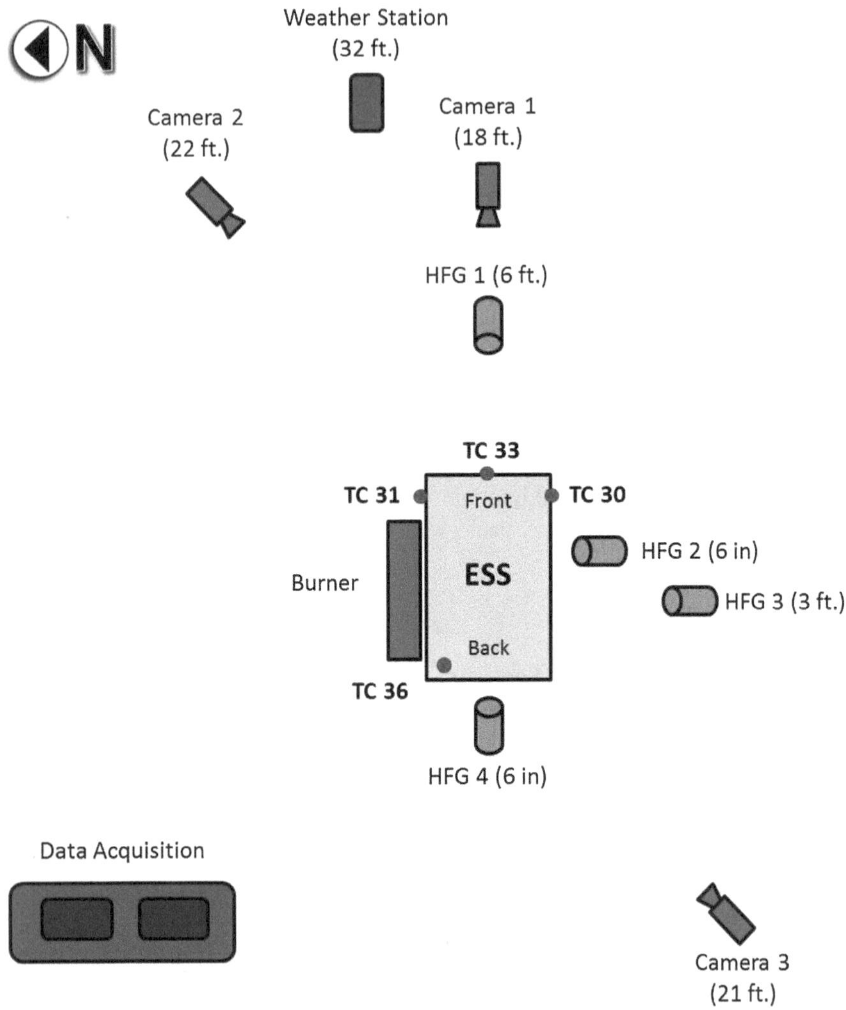

Fig. 5.1 External fire test setup and instrumentation

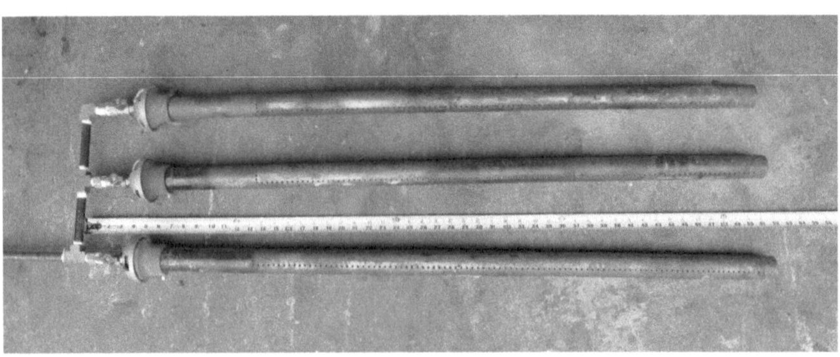

Fig. 5.2 Burners utilized for testing

Each burner was 1.25 in. in diameter and 49 in. long and consisted of two rows of 2 mm orifices drilled at an angle 45° apart. The orifices extended for 36 in., spaced approximately 5 mm apart. The three burners were installed inside a five sided, 36 by 36 in. by 8 in. steel shell, with a steel mesh covering the opening to increase the amount of radiative heat load from the burner assembly to the exterior wall of the Powerpack enclosure. The burner assembly was positioned against the right side of the Powerpack cabinet to allow for direct flame impingement on the exterior of the unit, as shown in Fig. 5.3.

Fig. 5.3 Burner assembly and positioning

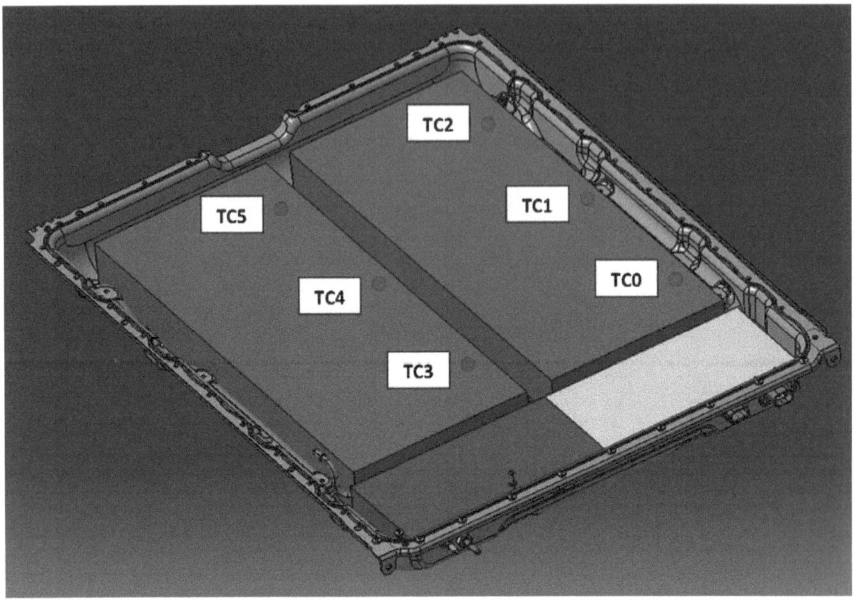

Fig. 5.4 TC measurement locations within pod 1; pods 2 through 4 are similarly instrumented and labeled in the same numerical order

The flow of propane was monitored by a calibrated Omega FMA1845A mass flow meter, capable of measuring up to 1000 liters per minute (lpm). The flow rate of propane was adjusted to provide an output of approximately 400 kW during the test.[1]

5.1.3 Temperature and Heat Flux Measurements

Temperatures were monitored with 1/8th-in. diameter bare bead Type K Chromel-Alumel thermocouples with an accuracy of ±2.2 °C or 0.75 %, whichever is greater. Twenty-nine (29) thermocouples were placed on the exterior surfaces of the Powerpack, at selected battery pods inside the Powerpack, and within the Powerpack cabinet and exhaust manifold. Six (6) thermocouples were installed inside pods 1, 2, 3, and 4, for a total of 24 thermocouples monitoring the thermal runaway progression inside the battery pods, as shown in Fig. 5.4. One (1) thermocouple was positioned inside the Powerpack cabinet exhaust manifold and another at the exhaust vent, as shown in Fig. 5.5. Three (3) additional

[1]HRR from the propane burners was determined based upon the flow rate of propane recorded by a mass flow meter during testing times the heat of combustion of propane.

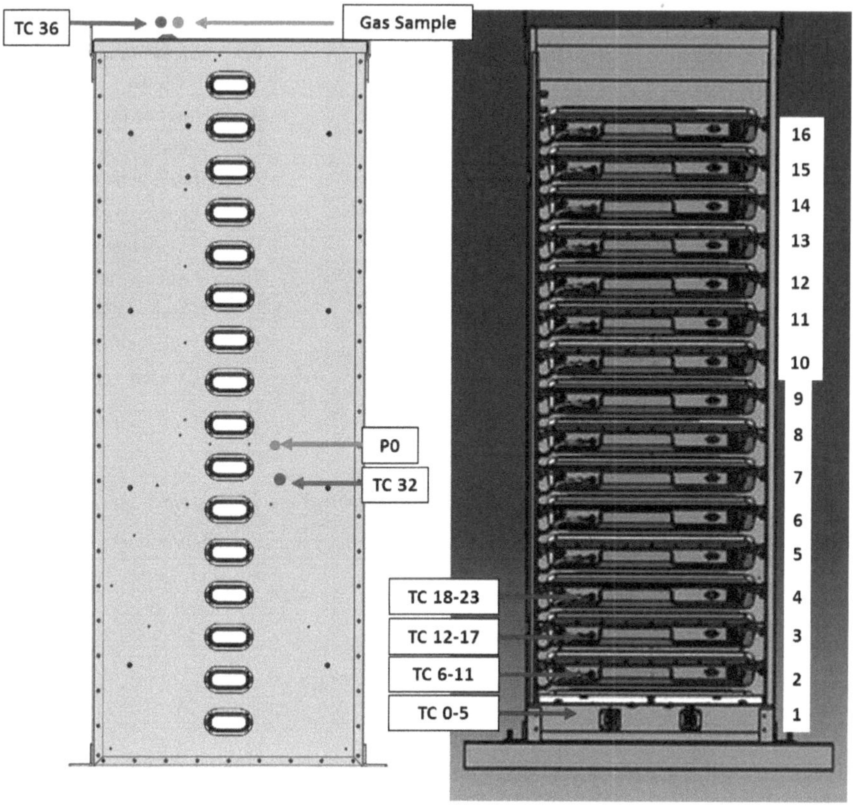

Fig. 5.5 TC, gas sampling, and pressure measurement locations inside the Powerpack cabinet

thermocouples were installed on the exterior surface of the Powerpack cabinet on the front, right side (burner side), and the left side of the Powerpack, as shown in Fig. 5.1.

Heat fluxes were monitored with Schmidt-Boelter heat flux gauges capable of measuring up to 50 kW/m^2 ± 3 %. The heat flux gauge has a target 0.60 in. in diameter that is enclosed within a water cooled body 2 in. in diameter. Four (4) heat flux gauges were placed 3 ft above the ground at standoff distances of 6 ft in front of the Powerpack, 6 in. and 3 ft from the left side (opposite of the burner) of the Powerpack, and 6 in. from the back of the Powerpack, as shown in Fig. 5.1. These distances are related to the clearance distances outlined in the Powerpack installation manual.

The location of each thermocouple and heat flux gauge is provided in Tables 5.1 and 5.2.

Table 5.1 Summary of thermocouple locations for external ignition testing

TC	Measurement location	TC	Measurement location	TC	Measurement location
0	Interior pod #1	10	Interior pod #2	20	Interior pod #4
1	Interior pod #1	11	Interior pod #2	21	Interior pod #4
2	Interior pod #1	12	Interior pod #3	22	Interior pod #4
3	Interior pod #1	13	Interior pod #3	23	Interior pod #4
4	Interior pod #1	14	Interior pod #3	30	Exterior left
5	Interior pod #1	15	Interior pod #3	31	Exterior right
6	Interior pod #2	16	Interior pod #3	32	Exhaust manifold
7	Interior pod #2	17	Interior pod #3	33	Exterior front
8	Interior pod #2	18	Interior pod #4	36	Exhaust vent
9	Interior pod #2	19	Interior pod #4		

Table 5.2 Summary of heat flux gauge locations for external ignition testing

Heat flux gauge	Measurement location	Heat flux gauge	Measurement location
1	Front (6 ft)	3	Left (3 ft)
2	Left (6 in.)	4	Back (6 in.)

5.1.4 Pressure Measurements

Pressures were monitored inside the Powerpack cabinet exhaust manifold using an Omega PX-309-015G5V pressure transducer capable of measuring up to 15 psi with an accuracy of ±0.25 % full-scale. The transducer was positioned inside the Powerpack cabinet exhaust manifold to detect any overpressures inside the Powerpack cabinet during the test, as shown in Fig. 5.5.

5.1.5 Products of Combustion Gas Sampling

Select products of combustion were monitored at the exhaust vent of the Powerpack cabinet as illustrated in Fig. 5.5. The gas samples were analyzed with a MultiRAE Lite PGM-6208 and a calibrated PortaSens II portable gas leak detector; model C16, manufactured by Analytical Technology, Inc. Gases measured included CO, chlorine (Cl_2), methane (CH_4) (monitored by the MultiRAE Lite) and HF (monitored by the PortaSens II) at a range up to 2000 ppm (±10 ppm), 50 ppm (±0.1 ppm), 0–100 % volume/volume (±0.1 %) and 100 ppm (±5 %), respectively. Previous experience with Li-ion battery fires and information provided by Tesla focused the gas analysis to these four gasses during this test series. The two detectors were portable handheld units that contained their own built in pumps to draw a gas sample from the exhaust vent through tubing into the respective detector chamber.

5.1.6 Weather Meter

A Kestrel 4500 weather meter was utilized to monitor the ambient temperature, humidity, wind speed, and direction during testing. The Kestrel was positioned approximately 32 ft away from the Powerpack in an open space, away from any structures or objects that could affect the conditions being monitored, as illustrated in Fig. 5.1.

5.1.7 Data Acquisition System

A National Instruments NI 9205 data acquisition unit was utilized to collect the heat flux and pressure measurements at a rate of 10 and 1000 measurements per second, respectively, at a 16 bit resolution. A MeasurePoint DT9874 Isolation Temperature data acquisition unit was utilized to collect temperature measurements at a rate of 10 measurements per second at a 24 bit resolution. The gas analyzers and the weather meter utilized their own built in data acquisition and recording software to collect data.

5.1.8 Still Photography and High Definition Video

Still images and high definition videos were taken throughout the test. Video cameras were positioned around the Powerpack to get a 360° view of the Powerpack at all times, as illustrated in Fig. 5.1. Still images were taken periodically during the test to capture the fire progression.

5.2 Internal Ignition Testing

The internal ignition test induced individual cells within the Powerpack to thermal runaway.

5.2.1 ESS Positioning

The Powerpack was positioned on a noncombustible surface similar to its intended end use installation on a concrete pad, as shown in Fig. 5.5. The test instrumentation, including thermocouples (TCs), pressure transducers, gas sampling, data

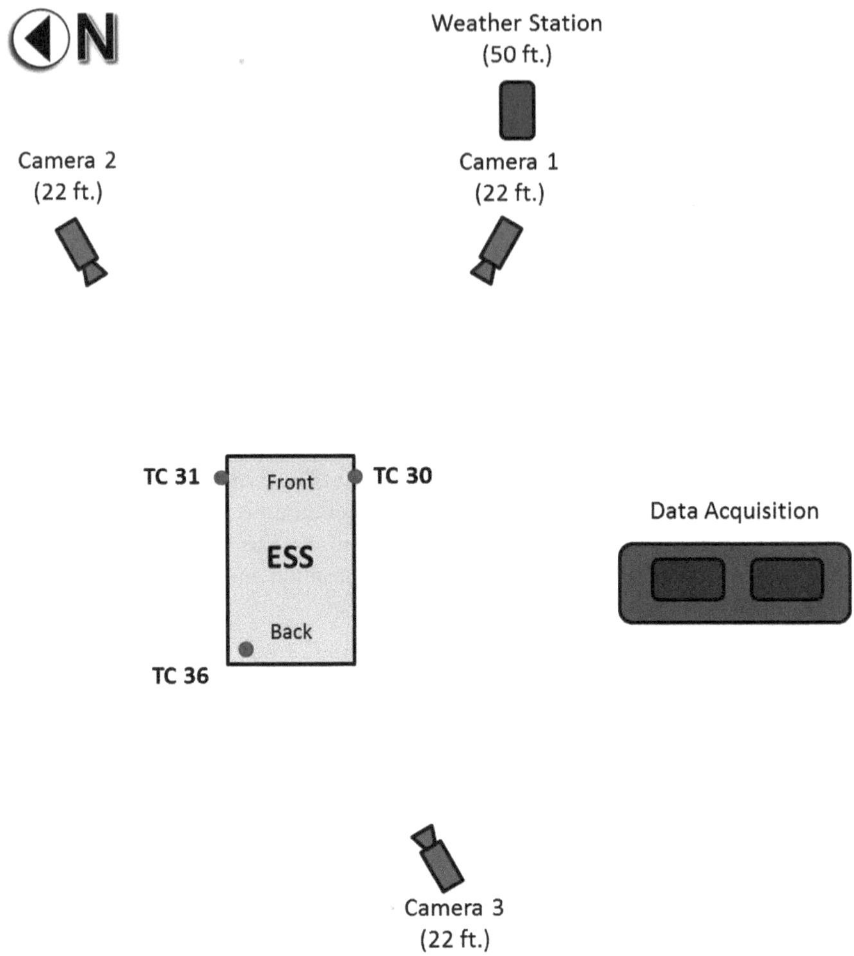

Fig. 5.6 Internal ignition test setup and instrumentation

acquisition, weather meter, and cameras were positioned around the Powerpack as illustrated in Fig. 5.6. In addition, heater cartridges utilized to force the individual batteries into thermal runaway, further described in Sect. 5.2.2, were positioned inside pod 6 (the initiator pod).

5.2.2 Internal Failure (Ignition Source)

The initiation method used in the internal ignition test consisted of using six (6) 1/8th-in. diameter 25-W cartridge heaters, each placed in an interstitial space between the battery cells in Pod 6, as shown in Fig. 5.8. All six heaters were clustered at the center of the module.

At the start of the test, current was applied to all six heaters simultaneously, resulting in an almost simultaneous thermal runaway of ten (10) cells. This method of inducing thermal runaway deliberately overwhelmed the passive propagation protection mechanisms of the Powerpack. After a minimum of ten cells had audibly undergone thermal runaway, the heaters were turned off.

5.2.3 Temperature Measurements

Temperatures were monitored with the same make and model 1/8th-in. diameter bare bead Type K Chromel-Alumel thermocouples as described in the external ignition testing. Thirty-seven (37) thermocouples were placed on the exterior surfaces of the Powerpack, at select battery pods inside the Powerpack, and within the Powerpack cabinet and exhaust manifold. Twelve (12) thermocouples were installed in the initiator pod (pod 6), as shown in Fig. 5.7 and six (6) thermocouples were placed inside pod 5 and pod 7, the adjacent pods to the initiator pod, as shown in Figs. 5.8 and 5.9. In addition, two (2) thermocouples were placed on the top cover of pod 6, two (2) on the bottom of the cover of pod 7 and two (2) on the top of the cover of pod 5, to monitor the spread of fire, if any, outside of the initiator pod, as shown in Fig. 5.10. Four (4) thermocouples were placed inside the

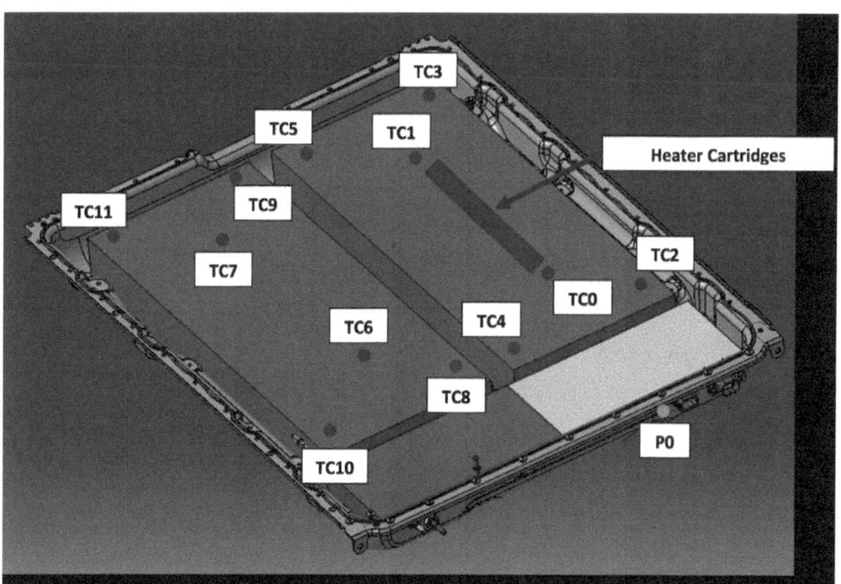

Fig. 5.7 TC, pressure measurement and heater cartridge locations within pod 6

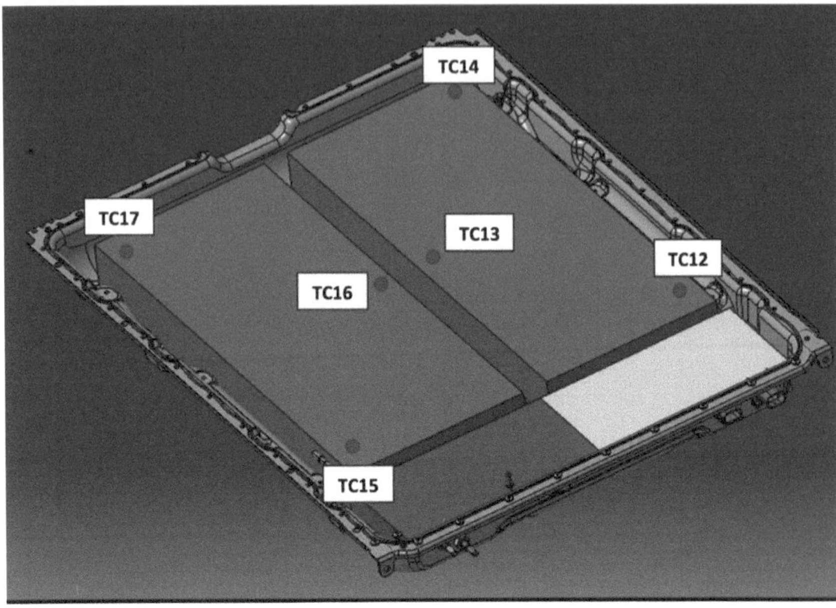

Fig. 5.8 TC measurement locations within pod 5

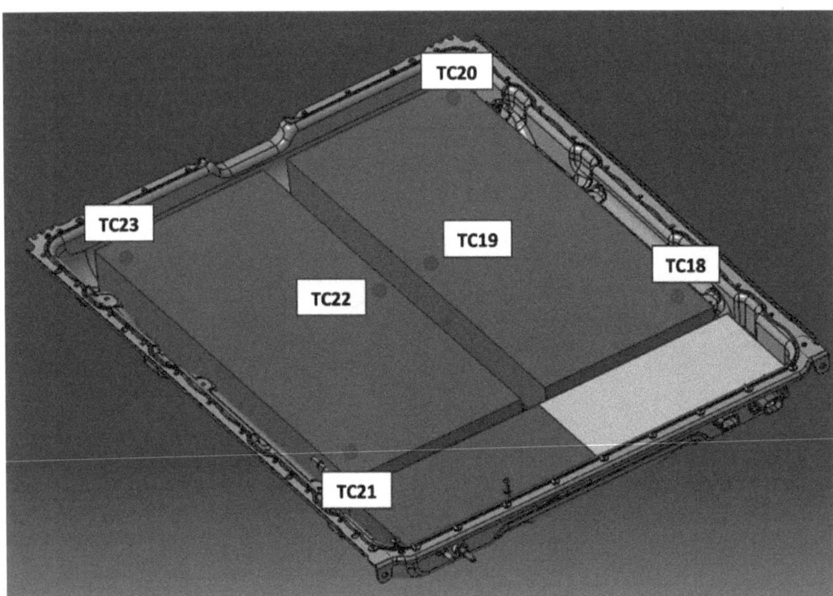

Fig. 5.9 TC measurement locations within pod 7

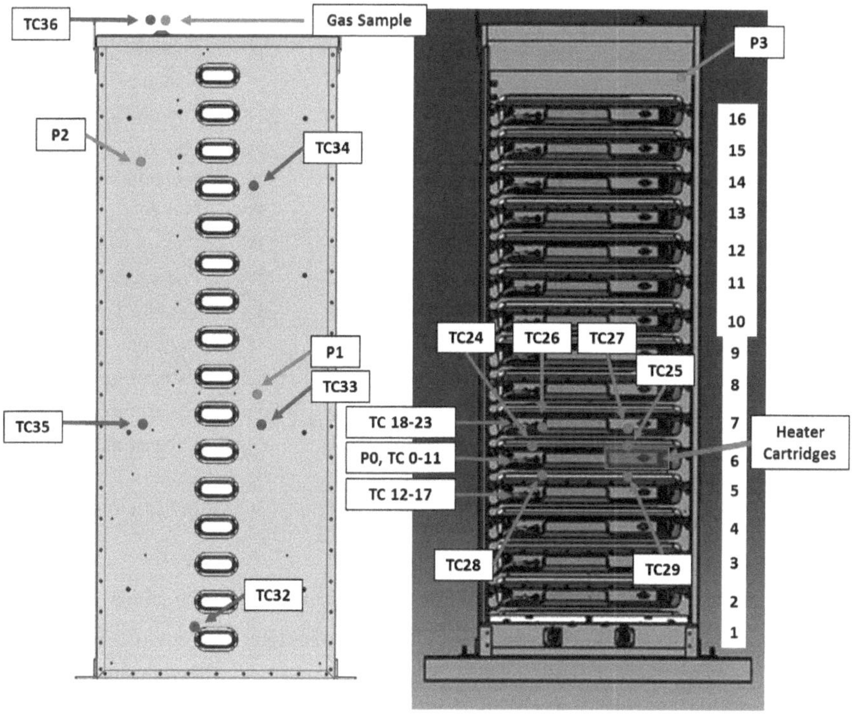

Fig. 5.10 TC, gas sampling, and pressure measurement locations inside the Powerpack cabinet

Powerpack cabinet in the exhaust manifold and another thermocouple was placed at the exhaust vent, as shown in Fig. 5.10. Two (2) final thermocouples were installed on the exterior surface of the Powerpack cabinet on the right and left sides of the Powerpack, as shown in Fig. 5.6. The location of each thermocouple is provided in Table 5.3.

5.2.4 Pressure Measurements

Pressures were monitored with the same make and model pressure transducers described in the external ignition testing. The transducers were positioned inside the Powerpack cabinet, as illustrated in Fig. 5.10, as well as in the exhaust manifold and at the initiator pod to detect any overpressures in these locations during the test. The location of the pressure transducers is provided in Table 5.4.

Table 5.3 Summary of thermocouple locations for internal ignition testing

TC	Measurement location	TC	Measurement location	TC	Measurement location
0	Interior pod #6	12	Interior pod #5	24	Pod #6 Cover
1	Interior pod #6	13	Interior pod #5	25	Pod #6 Cover
2	Interior pod #6	14	Interior pod #5	26	Pod #7 Cover
3	Interior pod #6	15	Interior pod #5	27	Pod #7 Cover
4	Interior pod #6	16	Interior pod #5	28	Pod #5 Cover
5	Interior pod #6	17	Interior pod #5	29	Pod #5 Cover
6	Interior pod #6	18	Interior pod #7	30	Exterior left
7	Interior pod #6	19	Interior pod #7	31	Exterior right
8	Interior pod #6	20	Interior pod #7	32	Exhaust manifold
9	Interior pod #6	21	Interior pod #7	33	Exhaust manifold
10	Interior pod #6	22	Interior pod #7	34	Exhaust manifold
11	Interior pod #6	23	Interior pod #7	35	Exhaust manifold
				36	Exhaust vent

Table 5.4 Summary of pressure measurement locations for internal ignition testing

Pressure transducer	Measurement location	Pressure transducer	measurement location
0	Pod #6	2	Exhaust manifold
1	Exhaust manifold	3	Front cabinet

5.2.5 *Products of Combustion Gas Sampling*

Select products of combustion were monitored at the exhaust vent of the Powerpack, as shown in Fig. 5.10, with the same instrumentation as described in the external ignition test (see Sect. 5.1.5).

5.2.6 *Weather Meter*

Weather conditions were monitored with the same instrumentation as described in the external ignition test (see Sect. 5.1.6). The weather meter was positioned approximately 50 ft away from the Powerpack, as shown in Fig. 5.6.

5.2.7 *Data Acquisition System*

The same data acquisition unit described in the external ignition test (see Sect. 5.1.7) was utilized to collect the test data during the internal ignition test.

5.2.8 Still Photography and High Definition Video

Still images and high definition videos were recorded throughout the internal ignition test. Video cameras were positioned around the Powerpack to get a 360° view of the Powerpack at all times, as illustrated in Fig. 5.6. Still images were taken periodically during the test to capture the test progression.

Chapter 6
Test Results

Exponent witnessed the full-scale testing and reviewed the data collected to observe the behavior of the Powerpack when it is involved in a fire scenario. The tests were performed at the Tesla test facility on November 5, 2015, under the guidance and direction of FPRF and Exponent. Two tests were conducted; one external ignition test and one internal ignition test. For each test the Powerpack was positioned out in open air, on a noncombustible surface, as it would be typically installed in outdoor installations. The Powerpacks that were tested were 100 kWh units charged to 100 % SOC, as described previously in Chap. 4.

6.1 External Ignition Testing

The external ignition test was conducted on November 5, 2015, at approximately 9:30 a.m. At the start of the test, the weather was overcast, with temperatures of approximately 35 °F and a relative humidity of approximately 65 %. The wind was out of the west-southwest with a wind speed of 1.5 miles per hour (mph). Over the course of the 3 h and 45 min test duration, the temperature slowly rose to 43 °F, the weather remained mostly overcast with no precipitation, the relative humidity dropped slowly to approximately 55 %, and the wind remained calm out of the west or west-southwest, with speeds between 0 and 2.2 mph. The following sections summarize the data collected during the test.

6.1.1 Test Observations

Table 6.1 summarizes the key events observed by Exponent during the test. Images at significant test times are provided in Figs. 6.1, 6.2, 6.3, 6.4, 6.5, 6.6, 6.7, and 6.8. In general, the test demonstrated that an external heat source, such as a propane

A.F. Blum and R.T Long Jr., *Fire Hazard Assessment of Lithium Ion Battery Energy Storage Systems*, SpringerBriefs in Fire,
DOI 10.1007/978-1-4939-6556-4_6

Table 6.1 External ignition test: key observations

Time (h:min:s)	Event
−0:03:00	Start data acquisition and video cameras
0:00:00	Ignite burner
0:35:12	First smoke (white and/or grey) observed from Powerpack
0:45:02	Pop sound heard from Powerpack cabinet (pops)
0:46:54	Sustained flames first observed at exhaust vent
0:47:09	Sustained flames first observed at back Powerpack panel
0:47:57	Sustained flames first observed at the front door
0:48:00	Steady pops heard from Powerpack starting at this time until 3:10:50 consistent with cell thermal runaway
1:00:00	Burners "OFF", jet fire exiting exhaust vent, flames coming out of the front door grill
1:05:00–1:10:00	Material ejected from exhaust vent
1:08:00	Fire inside Powerpack only involving combustibles near the top; no burning of materials near the bottom of the Powerpack
1:11:00	Jet flame at exhaust vent weakens intermittently
1:14:30	Jet flame at exhaust vent increases in intensity
1:20:05	Overpressure sound consistent with refrigerant failure
1:27:00	Jet flame at exhaust vent weakens intermittently
1:27:45	Smoke from Powerpack turns darker
1:29:45	Flames inside Powerpack moving lower
1:32:05	Fire inside Powerpack intensifying
2:00:00	Fire inside Powerpack intensifying
2:30:00	Fire inside Powerpack steady
2:33:30	Fire inside Powerpack decreasing in intensity
3:00:00	Fire insides subsiding, weak flames observed out the front door and exhaust vent
3:10:50	Last pop heard from Powerpack
3:30:00	Fire continues to decrease in intensity
3:41:10	Last visible flame out
3:45:00	Cameras and data acquisition off

burner, could induce the Powerpack into thermal runaway and result in the ignition of electrolyte material and other combustibles within the Powerpack cabinet. Popping sounds from the interior of the Powerpack were heard throughout the test. White smoke was observed consistent with the release of flammable electrolyte material from individual cells. However, no violent projectiles, explosions, or bursts (other than an overpressure release of the thermal door refrigerant) were observed during the test while the Powerpack was exposed to the burners, while it was in a

Fig. 6.1 External test screenshot: start of test, burners "ON"

Fig. 6.2 External test screenshot: test time = 1 h, fire emanating from the front door and exhaust vent, pops consistent with battery cell thermal runaway heard steadily, burners turned "OFF"

free burn state, or after flames were no longer visible. Flames remained mostly confined to the Powerpack itself. Weaker flames emanated from the exhaust vent of the Powerpack, the front thermal door grill, and around the front thermal door seal at varying times throughout the test.

Fig. 6.3 External test screenshot: test time = 1 h 30 min, flames at front door and exhaust vent intermittently decreasing in intensity

Fig. 6.4 External test screenshot: test time = 2 h, fire inside the Powerpack intensifying

6.1.2 Temperatures and Heat Flux Measurements

Temperature and heat flux measurements were collected during the external ignition test and plots for each as a function of time are provided in Appendix A.

The maximum temperatures measured on the interior thermocouples installed within pods 1, 2, 3, and 4 were all consistent with direct flame contact, with temperatures in excess of 2000 °F. The maximum temperatures measured on the

Fig. 6.5 External test screenshot: test time = 2 h 30 min, fire inside the Powerpack intensifying

Fig. 6.6 External test screenshot: test time = 3 h, fire inside the Powerpack subsiding

exterior of the Powerpack cabinet were much lower. TC30, positioned on the left side of the Powerpack opposite of the burner, measured a maximum surface temperature of approximately 150 °F and TC33, positioned on the front door measured a maximum temperature of approximately 460 °F.

An analysis of the heat flux measurements yielded values inconsistent with observations of the test, the fire progression and its severity. As such, the data collected for heat fluxes was not considered in this fire hazard assessment.

Fig. 6.7 External test screenshot: test time = 3 h 30 min, fire inside the Powerpack burning itself out

Fig. 6.8 External test screenshot: end of test, fire is out

6.1.3 Pressure Measurements

Pressure was monitored at the Powerpack exhaust manifold throughout the test and a plot of the pressure as a function of time is provided in Appendix B. No pressure build-up or release consistent with an overpressure event occurring inside the Powerpack cabinet was observed in the data or during the test.

6.1.4 *Gas Sampling Measurements*

Select products of combustion were monitored at the Powerpack exhaust vent throughout the test and a plot of CO and HF levels as a function of time are provided in Appendix C.

CO was first detected approximately 2.5 min after the burners were turned on at 10 ppm. The value steadily rose to its maximum value of 50 ppm approximately 4 min after the burners were turned on. The CO detected then slowly decreased to 0 ppm approximately 30 min after the burners were turned on and remained at 0 ppm for the remainder of the test. As such, the production of CO, as detected at the exhaust vent, only occurred while the external burner was on and CO was not detected while the Powerpack underwent self-sustaining combustion (i.e., the external burner was off).

No Cl_2 or CH_4 were detected in any quantities during the test.

HF was detected 2 min after the burners were turned on at 2 ppm. The value of HF steadily rose from 2 ppm to its maximum value of 100 ppm approximately 30 min after the burners were turned on. The maximum range of the HF detector was 100 ppm. All HF data after 30 min was "over range" of the HF detector, indicating HF levels were greater than 100 ppm for the duration of the test.

6.1.5 *Post Test*

Following the test, it was determined that all of the energy pods were damaged and there was no stranded energy within the Powerpack.

6.2 Internal Ignition Testing

The internal ignition test was conducted on November 5, 2015, at approximately 2:45 p.m. At the start of the test, the weather was sunny, with temperatures of approximately 49 °F and a relative humidity of approximately 32 %. The wind was out of the west with a wind speed of 1.3 mph. Over the course of the hour and a half test duration, the temperatures fluctuated between approximately 45 and 54 °F, the weather remained mostly sunny with no precipitation, the relative humidity remained between 30 and 32 %, and the wind remained calm out of the west or west-southwest, with speeds between 0.1 and 2.5 mph. The following sections summarize the data collected during the test.

6.2.1 Test Observations

Table 6.2 summarizes the key events observed during the test. Images at significant test times are provided in Figs. 6.9, 6.10, and 6.11. In general, the internal ignition test demonstrated that heater cartridges installed within the battery pack could induce multiple battery cells into thermal runaway; however, the failures did not result in thermal runaway of battery cells outside of the initiator pod. Popping sounds from the interior of the Powerpack were heard sporadically throughout the test, and steadily for approximately 15 min. White smoke was observed consistent with the release of flammable electrolyte material from individual cells. However, no violent projectiles, explosions, or bursts were observed during the test. In addition, no flames or other signs of fire, other than smoke production, were observed. The event stopped on its own without thermal runaway occurring outside of the initiator pod.

Table 6.2 Internal ignition test: key observations

Time (h:min:s)	Event
−0:01:30	Start data acquisition and video cameras
0:00:00	Turn on heater cartridges
0:12:35	Pop sound heard from Powerpack cabinet (pops)
0:15:10	First smoke (white and/or light grey) observed at exhaust vent
0:27:13	Light smoke continues at exhaust vent
0:29:35	Smoke at exhaust vent increasing
0:33:07	Pop heard from Powerpack cabinet
0:34:28	Smoke at exhaust vent increasing, getting darker (grey)
0:34:56	Pop heard from Powerpack cabinet
0:35:30	Smoke at exhaust vent increasing, getting darker (grey)
0:36:22	Steady pops heard from Powerpack starting at this time until 0:45:01 consistent with cell thermal runaway
0:38:34	Heater cartridges turned off
0:45:01	Last pop heard from Powerpack
0:49:30	Smoke production at exhaust vent subsiding
1:00:00	Smoke production at exhaust vent subsiding
1:15:00	Smoke production at exhaust vent subsiding
1:30:00	Smoke production at exhaust vent barely visible, cameras and data acquisition turned off, test terminated

Fig. 6.9 Internal test screenshot: start of test, heater cartridges "ON"

Fig. 6.10 Internal test screenshot: peak smoke production approximately 35–40 min after the heater cartridges were turned "ON"

6.2.2 Temperatures

Temperatures were collected during the internal ignition test and plots of the temperatures as a function of time are provided in Appendix D.

The maximum temperatures were measured in the initiator pod, Pod 6. In the module with the heater cartridge, maximum temperatures were recorded in excess of 2000 °F for approximately 2 s at TC1, which was one of the thermocouples installed closest to the heater cartridges and may have come in contact with a brief

Fig. 6.11 Internal test screenshot: end of test

(2 s) flame as the cell underwent thermal runaway. The rest of the thermocouples within the module recorded temperatures up to approximately 1550 °F. In the second module within pod 6, the temperatures were lower, with the maximum temperatures between 200 and 400 °F for the six thermocouples installed within that module. The maximum temperatures measured on the adjacent pods, pod 5 and pod 7, were much lower as well. Pod 5 recorded maximum temperatures between 80 and 125 °F and pod 7 recorded maximum temperatures between 80 and 180 °F. TC30, positioned on the left side of the Powerpack cabinet exterior, measured a maximum surface temperature of 70 °F and TC31, positioned on the right side of the Powerpack cabinet exterior, measured a maximum temperature of 60 °F.

6.2.3 Pressure Measurements

Pressure was monitored inside the Powerpack cabinet, exhaust manifold and at the initiator pod throughout the test and a plot of the pressure as a function of time is provided in Appendix E. No pressure build-up or release consistent with an overpressure event occurring inside the Powerpack cabinet or the initiator pod was observed in the data or during the test.

6.2.4 Gas Sampling Measurements

Select products of combustion were monitored at the Powerpack exhaust vent throughout the test and a plot of CO, CH_4, and HF levels as a function of time is provided in Appendix F.

CO was first detected approximately 10.5 min after the heaters were turned on at 10 ppm. The value steadily rose to its maximum value of 2000 ppm approximately 12 min after the heaters were turned on, which is the maximum range for the CO detector. The value of CO remained at its maximum detection level of 2000 ppm from the 12 min mark until 63.5 min after the heaters were turned on. It then slowly decreased for the remaining 30 min of the test.

No Cl_2 was detected in any quantities during the test.

CH_4 was first detected approximately 12 min after the heaters were turned on. The detector measured CH_4 in percent volume fraction and steadily rose until approximately 36 min after the heaters were turned on, to a recorded a maximum percentage of 96.9. This time correlates with when the most cell runaways were observed in the test, as described in Sect. 6.2.1. It then slowly decreased for the remaining 54 min of the test.[1] The elevated CO and CH_4 levels detected after the heater cartridges were turned off and after thermal runaway of the cells had ceased indicates that CO and CH_4 can still be vented from the cells as they are cooling and obvious signs of thermal runaway (i.e., popping) are no longer observed.

HF was detected approximately 21 min after the heater cartridges were turned on at 1 ppm. The value of HF steadily rose from 1 ppm to its maximum value of 26 ppm approximately 46 min after the heater cartridges were turned on. The value plateaued at 26 ppm for 2 additional minutes (minute 47 and 48), then steadily declined back down to a value of 2 ppm by the end of the test.

6.2.5 *Post Test*

Following the test, it was determined that only one of the energy pods (the initiator pod) was damaged. The other 15 pods remained operational and had a full SOC. The energy pods were discharged and the Powerpack was recycled.

[1]During thermal runaway of the battery cell methane can be released. During the external ignition test no methane was detected at the exhaust vent, likely a result of the fire inside the ESS igniting any off gassing methane from the cells. However during the internal ignition test, no flames were observed and the released methane vented into the exhaust manifold and out the exhaust vent. Methane was also detected in previous testing programs, such as during the FAA's fire tests of cylindrical battery cells.

Chapter 7
Key Findings

This chapter is a discussion of the data and observations collected during the literature review and full-scale testing and supplements the presentation of the data in Chaps. 2 and 6.

7.1 Literature Review Summary

Li-ion ESSs are becoming more popular and are posed to be installed in many occupancies across the country, including commercial and residential buildings. However, little public knowledge is known about the fire hazards they pose to those buildings and their occupants.

7.1.1 Electrical, Fire, and Building Codes

Several gaps were identified in a review of electrical, fire, and building codes typically adopted in the United States as they relate to ESSs. These gaps are predominantly related to sections of the codes categorizing battery systems based on the volume of liquid electrolyte, which is not appropriate for assessing Li-ion ESS hazards. In addition, NFPA 1 provides contradictory guidance regarding thermal runaway protection for Li-ion battery systems, while the IFC does not require it at all. These gaps can be corrected with changes to the sections identified at the ICC code action hearings and NFPA technical committee meetings, some of which are currently being undertaken. In addition, the next edition of the NEC, the 2017 edition, is proposed to have a new article (Article 706) dedicated to ESSs. This addition should further assist installers, AHJs, and manufacturers with navigating the electrical installation requirements for these systems. However, it should also be noted that guidance for Li-ion battery system installations is currently within

A.F. Blum and R.T Long Jr., *Fire Hazard Assessment of Lithium Ion Battery Energy Storage Systems*, SpringerBriefs in Fire,
DOI 10.1007/978-1-4939-6556-4_7

the codes and has been since 2006, most notably Section 608 of the IFC and since 2009, Chapter 52 of NFPA 1. Many of the concerns over the installation of battery systems could be addressed by local jurisdictions adopting more current editions of the ICC codes.

7.1.2 Design Standards

The ESS assessed in this testing program was listed to UL 1741, UL 1973, and IEC 62109 and was designed to be compliant with UL 9540 and IEC 62619, currently under development. UL 1973 requires stationary battery systems to meet two fire tests: one originating internally at the battery cell level and one externally by means of a hydrocarbon pool fire.

7.1.3 ESS Fires

Real world experience with Li-ion ESS fire incidents are limited, likely stemming from the early stage of adoption that these systems are currently in. Only one case was identified in the public records where a Li-ion ESS was involved in a fire; however, the details of that fire are not known, as requests for more information for public sources have not yielded any additional details. Previous research on other large format Li-ion batteries had demonstrated that the batteries did not significantly add to the HRR of the fire, that the fires can be extinguished with large amounts of water, the batteries can pose a projectile hazard when designed with cylindrical 18,650 cells, but do not pose that hazard with polymer or pouch style cells, that toxic compounds such as CO_2, NO_x, HCN, HCl, CO, and HF can be produced during the fires, water samples collected after extinguishing Li-ion battery fires can contain concentrations of fluoride and chloride, and that no electrical hazards exist for personnel suppressing a battery fire from current leakage through the hose stream provided they are standing at specified standoff distances.

7.1.4 Knowledge Gaps

As stated in Chap. 2, the following gaps in the knowledge base for commercial and residential Li-ion ESSs have been identified

1. No public fire test data demonstrating the fire behavior of ESSs.
2. Limited public fire test data related to large format battery packs with cylindrical design utilized either in vehicles or storage systems.

3. No fire test data or real world fire incidents involving residential or commercial Li-ion ESSs illustrating the hazards (projectiles, heat release, toxic gas production) to first responders and/or the best practices for fire department operations.
4. No Li-ion ESS guidance in the IRC.
5. Limited real world fire incidents involving large-scale (grid size) ESSs.
6. Some sections of the IBC, IFC, and NFPA 1 are confusing, as only the volume of the electrolyte (a requirement for older battery chemistries such as lead acid) and not the weight of the Li-ion battery system, is used as a threshold for when certain building or fire code requirements are necessary. In addition, other agencies, such as the United Nations and DOT, have other methods for defining and categorizing batteries. Many of these code sections are presently being revised and could be addressed by the next published code set.
7. NFPA 1 provides contradictory guidance regarding thermal runway protection for Li-ion battery systems, while the IFC does not require thermal runaway protection for Li-ion battery systems at all. Many of these code sections are presently being addressed and could be resolved by the next published code set.
8. No post-fire incident response and recovery (i.e., overhaul) procedures.
9. No stationary battery system or ESS fire reporting code in NFIRS to assist in analyzing fire incidents and differentiate battery systems from household batteries.

7.2 Test Summary

The following sections highlight the key findings from the full-scale fire tests.

7.2.1 Overall Test Observations

A 400 kW propane burner impinging directly on the side of the Powerpack for approximately 60 min was required to achieve self-sustaining thermal runaway in the Powerpack battery pack and ignite interior components within the Powerpack cabinet. The test had a duration of approximately 3 h and 45 min until the fire burned itself out. Flames were observed breaching the cabinet at the front door of the Powerpack and out the top of the Powerpack at the exhaust vent. No projectiles or explosions were observed at any time during either test.

During the internal ignition test, individual battery cells were forced into thermal runaway; however, no flames were observed at any time. Smoke was observed emanating from the Powerpack at the exhaust vent, however, within 1 h and 30 min the smoke had dissipated and the thermal event was over. The Powerpack was designed to stop a single battery cell failure from cascading into a series of thermal runaways of adjacent battery cells, a design safety feature deliberately overwhelmed in this test through the use of multiple heater cartridges. However, the event was still contained within the Powerpack and did not propagate outside of the initial pod where the heaters were installed.

7.2.2 Flame Spread Hazards

Temperature measurements in the external ignition test demonstrated that a fire inside the Powerpack can reach elevated temperatures in excess of 2000 °F. Exterior temperatures at the Powerpack cabinet were much lower and would not pose a fire spread hazard if the manufacturer recommended clearance distances to combustibles, as specified by the installation manual, are followed. Flames did breach the front door; however, the recommended clearance distance of six feet would likely eliminate any direct flame spread from the front door to nearby combustibles. Given that the unit tested can be installed outdoors, wind conditions could affect any flames emanating from the Powerpack. During these tests, the wind was calm with speeds at or less than 2 mph. As such, the hazard that a high wind scenario could inflict on the flame spread was not directly assessed during these two tests and may warrant further investigation. In addition, a standalone Powerpack was tested in this test program, not a large installation with many Powerpacks installed in an array. As such, the effects, if any, of additional Powerpacks installed within close proximity to one another was not directly assessed during these two tests and may warrant further investigation.

Flames several feet high were observed from the exhaust vent at the top of the Powerpack. The installation manual recommends at least five feet of clearance above the Powerpack. This clearance may not be sufficient if combustible materials are installed above the Powerpack, such as a building canopy or awning. It is recommended that this clearance distance be evaluated when a system is being installed, especially if the installed system is adjacent to a building or structure that has or could have combustibles installed above the Powerpack.

During the internal ignition test the temperatures recorded were much lower, with exterior cabinet surface temperatures only slightly higher than ambient and no observed flames emanating from inside the Powerpack. Based on this test, the flame spread hazard from an internal cell failure for combustibles positioned at the recommended clearance distances away from the Powerpack is negligible.

7.2.3 Products of Combustion Hazards

The release of HF during Li-ion fires is well known and HF was detected in both fire tests. The maximum range for the portable detector utilized in testing was 100 ppm, which was exceeded during the external ignition test after 30 min of burner exposure to the Powerpack. During the internal ignition test, the maximum recorded HF was 26 ppm, as less battery cells were involved compared to the external ignition test. Both of these measurements are greater than the recommended exposure levels over an 8 h period as specified by the Occupational Safety and Health Administration (OSHA). It is recommended that first responders don typical firefighting self-contained breathing apparatus (SCBA) equipment when responding to an outdoor Li-ion battery fire. CO was also detected in both fire tests, though more significantly in the internal ignition fire test. Based on these test results, if installed indoors, additional ventilation of the Powerpack and/or for the room in which it is installed may be required. In addition, this test series only assessed select products of combustion produced during the Powerpack fires, namely HF. Additional testing accounting for other toxic products of combustion may warrant further investigation.

Chapter 8
Recommendations and Future Work

The following recommendations and possible future work are suggested (Phase II) to further identify and understand the fire hazards of Li-ion ESSs:

- Research studying first responder tactics and suppression for Li-ion ESS fires.
- Research studying post fire incident response and recovery (i.e., overhaul) procedures.
- Heat release rate testing of ESSs.
- Testing to study what effect, if any, severe wind conditions may have on the spread of flames from one ESS to another or to other nearby combustibles.
- Testing to study what effect, if any, an array of ESSs installed within close proximity to one another would have on the spread of flames from one ESS to another or to other nearby combustibles.
- Testing of ESSs inside a compartment to study what effect, if any, a room will have on the fire behavior and potential toxic gas hazards within an enclosure.
- Testing to study different ESS manufacturers' products, battery chemistries, and/or sizes under similar conditions to verify the performance of other ESSs under these fire conditions.
- The addition of a stationary battery or ESS code in NFIRS such that fires in these systems can be differentiated from other battery fires, such as household batteries.
- Resolve the conflicting code sections relating to ESSs.

A.F. Blum and R.T Long Jr., *Fire Hazard Assessment of Lithium Ion Battery Energy Storage Systems*, SpringerBriefs in Fire,
DOI 10.1007/978-1-4939-6556-4_8

Appendix A
External Ignition Test: Temperature and Heat Flux Plots

See Figs. A.1, A.2, A.3, A.4, A.5, A.6, A.7, A.8, A.9, and A.10.

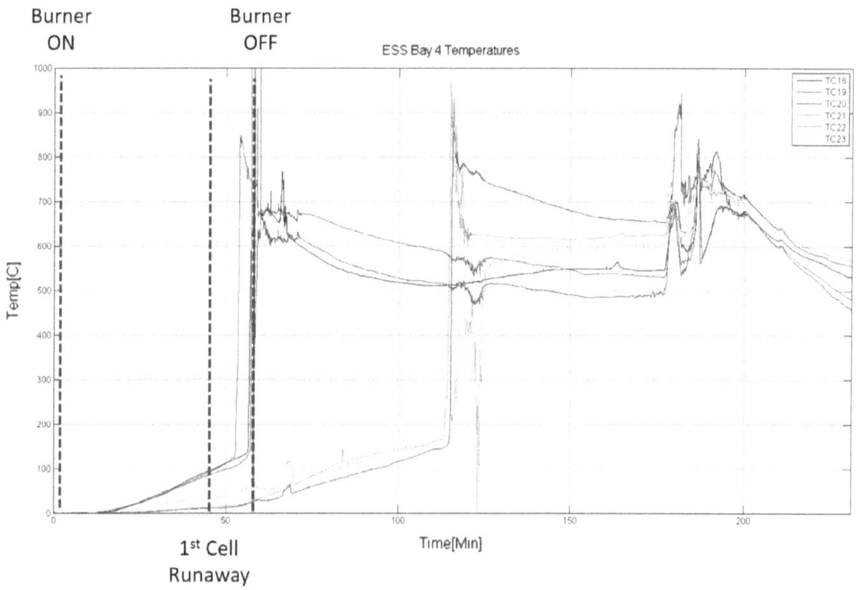

Fig. A.1 Powerpack pod 4 temperatures (noise observed in the data is consistent with electrical interference that occurs during voltage leakage from the damaged batteries after thermal runaway)

A.F. Blum and R.T Long Jr., *Fire Hazard Assessment of Lithium Ion Battery Energy Storage Systems*, SpringerBriefs in Fire,
DOI 10.1007/978-1-4939-6556-4

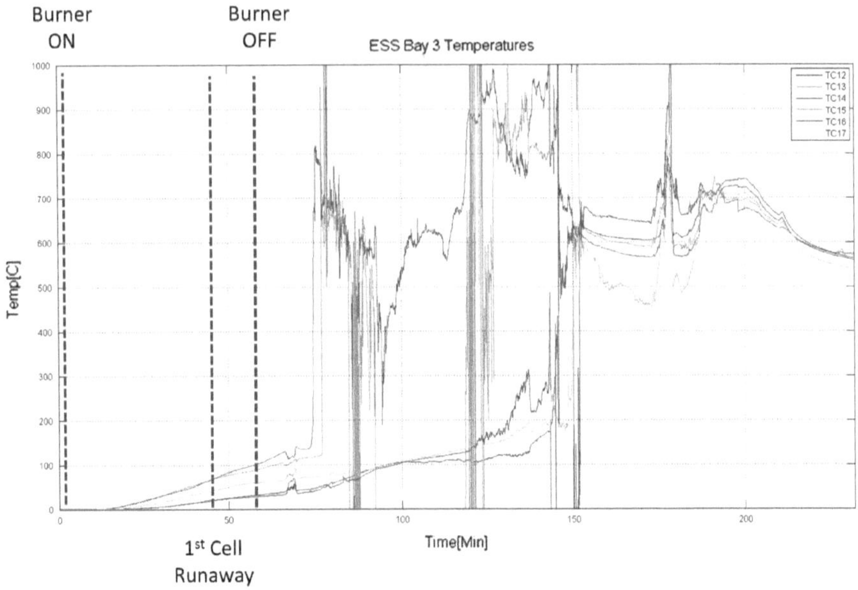

Fig. A.2 Powerpack pod 3 temperatures (noise observed in the data is consistent with electrical interference that occurs during voltage leakage from the damaged batteries after thermal runaway)

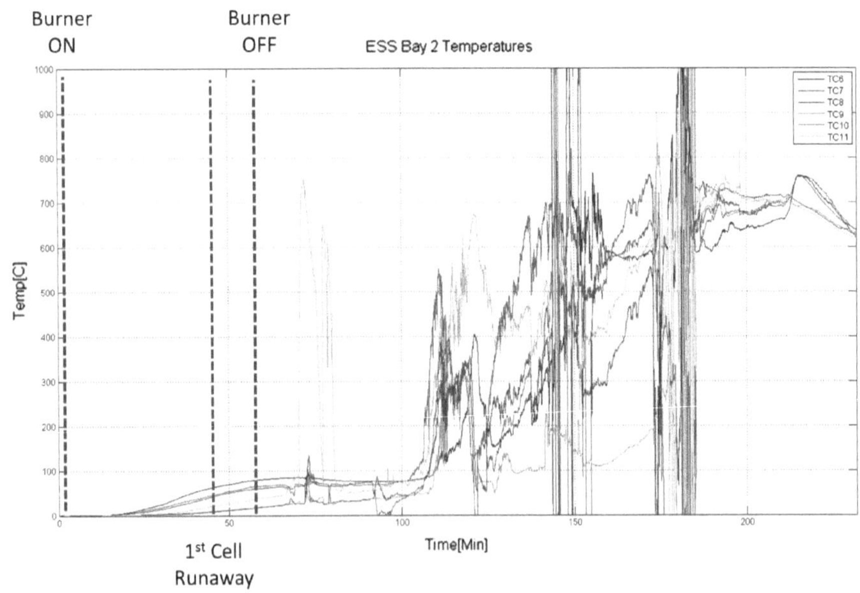

Fig. A.3 Powerpack pod 2 temperatures (noise observed in the data is consistent with electrical interference that occurs during voltage leakage from the damaged batteries after thermal runaway)

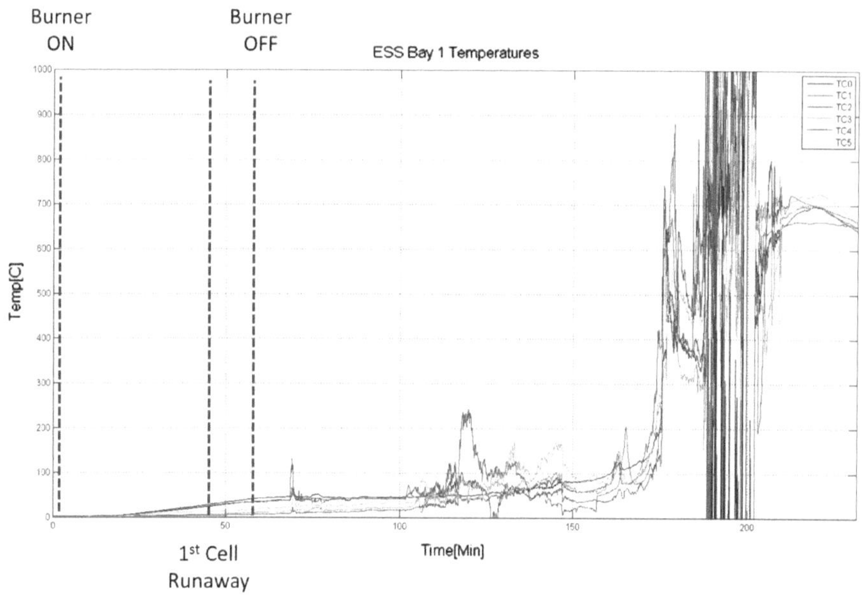

Fig. A.4 Powerpack pod 1 temperatures (noise observed in the data is consistent with electrical interference that occurs during voltage leakage from the damaged batteries after thermal runaway)

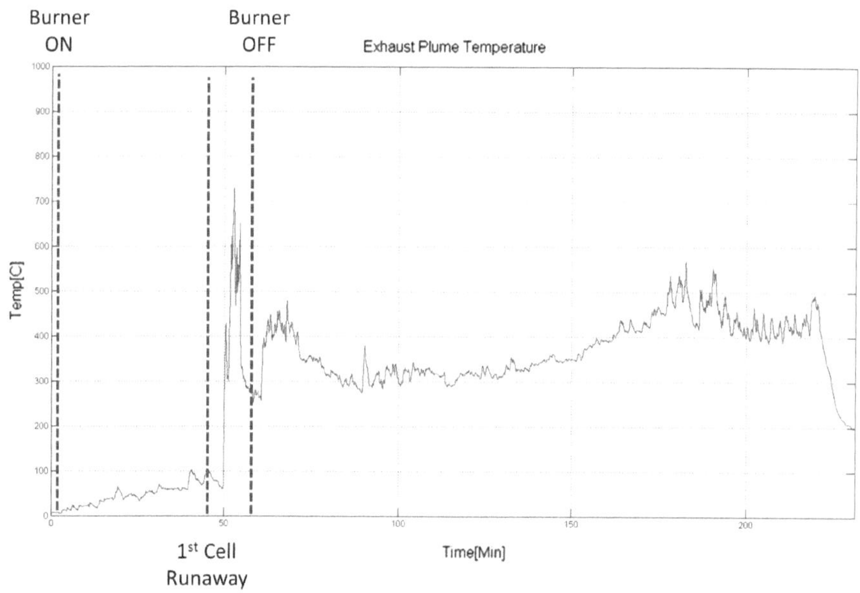

Fig. A.5 Powerpack exhaust vent temperature

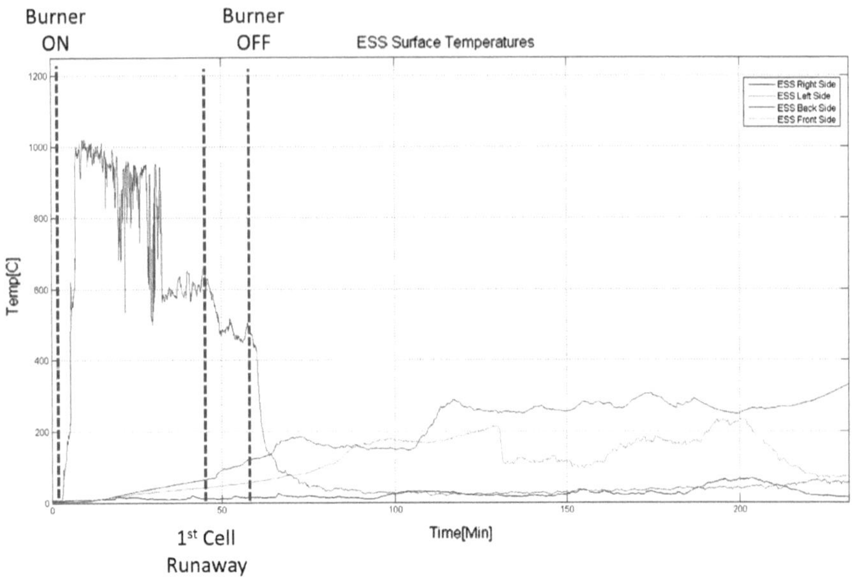

Fig. A.6 Powerpack external surface temperatures

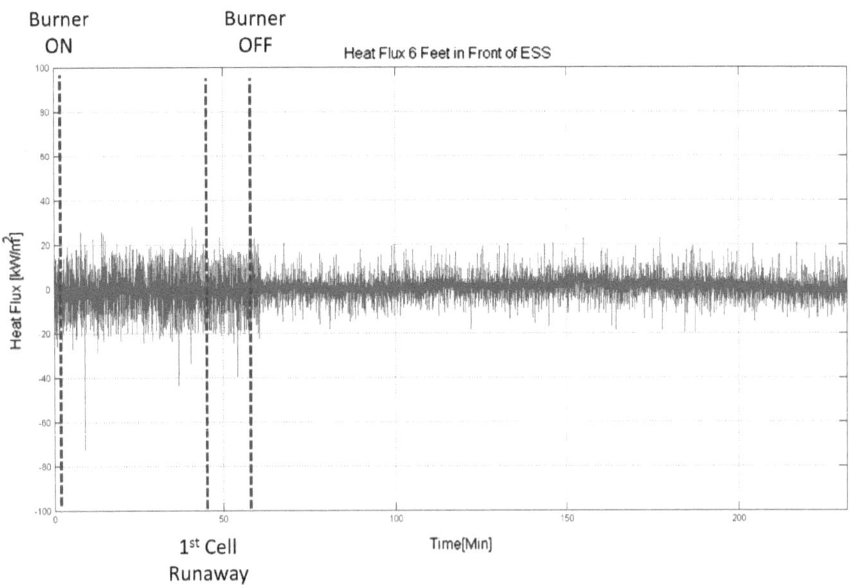

Fig. A.7 Heat flux measurements at HFG1, 6 ft from the front of the Powerpack

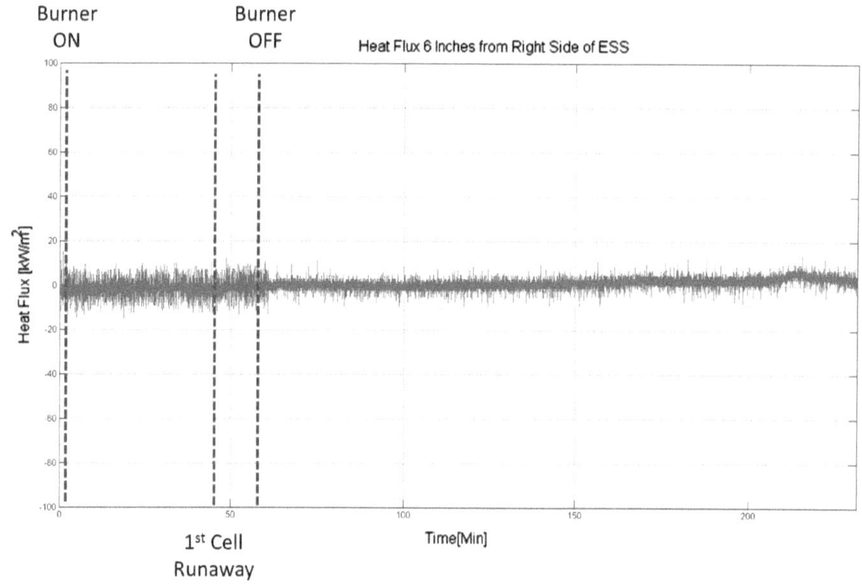

Fig. A.8 Heat flux measurements at HFG2, 6 in. from the side of the Powerpack

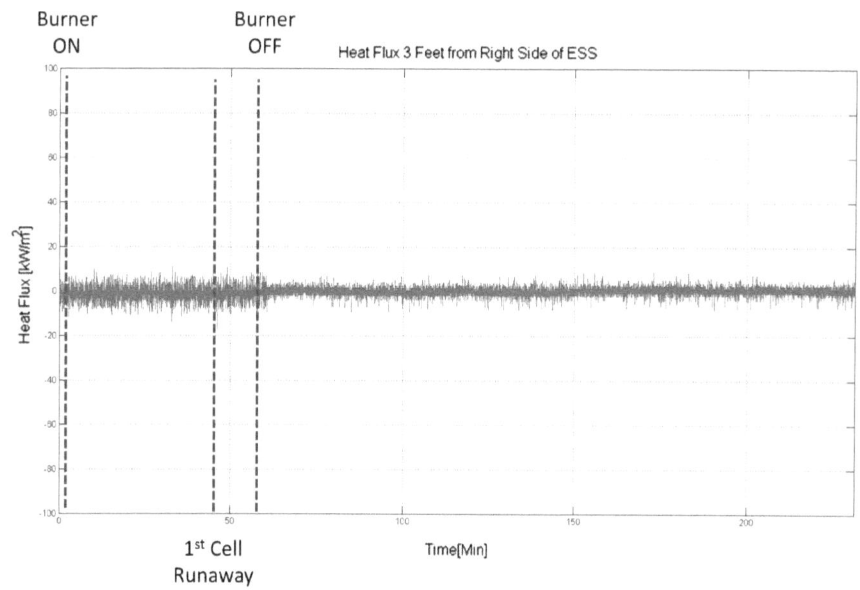

Fig. A.9 Heat flux measurements at HFG3, 3 ft from the back of the Powerpack

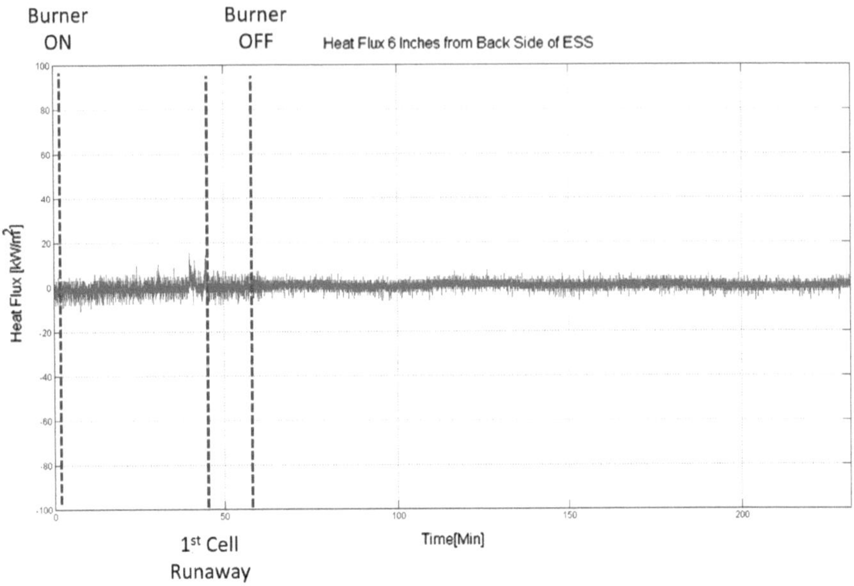

Fig. A.10 Heat flux measurements at HFG4, 6 in. from the side of the Powerpack

Appendix B
External Ignition Test: Pressure Plot

See Fig. B.1.

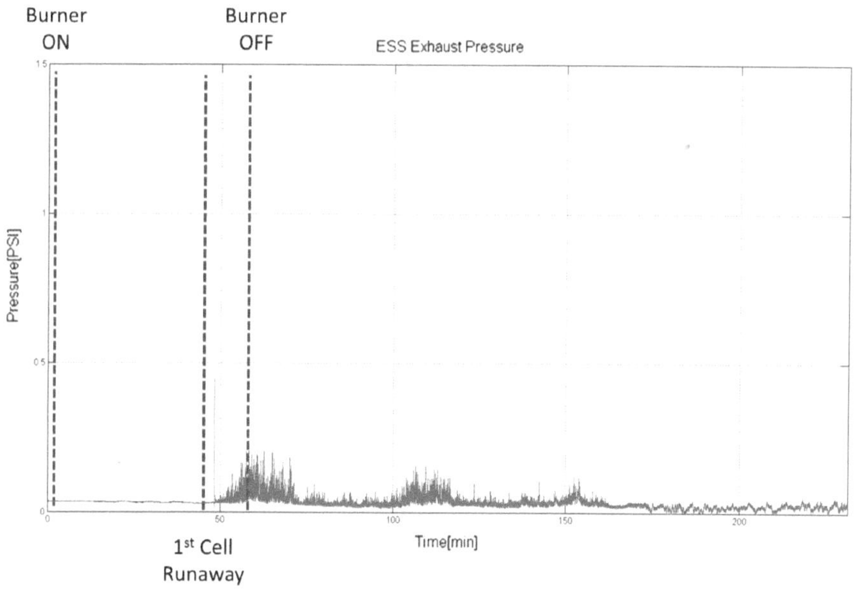

Fig. B.1 Exhaust manifold pressure

A.F. Blum and R.T Long Jr., *Fire Hazard Assessment of Lithium Ion Battery Energy Storage Systems*, SpringerBriefs in Fire,
DOI 10.1007/978-1-4939-6556-4

Appendix C
External Ignition Test: Gas Sampling Plot

See Figs. C.1 and C.2.

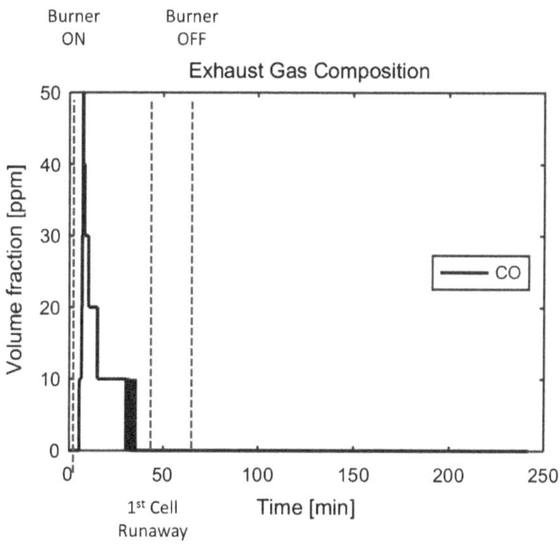

Fig. C.1 CO detected at the exhaust vent

A.F. Blum and R.T Long Jr., *Fire Hazard Assessment of Lithium Ion Battery Energy Storage Systems*, SpringerBriefs in Fire,
DOI 10.1007/978-1-4939-6556-4

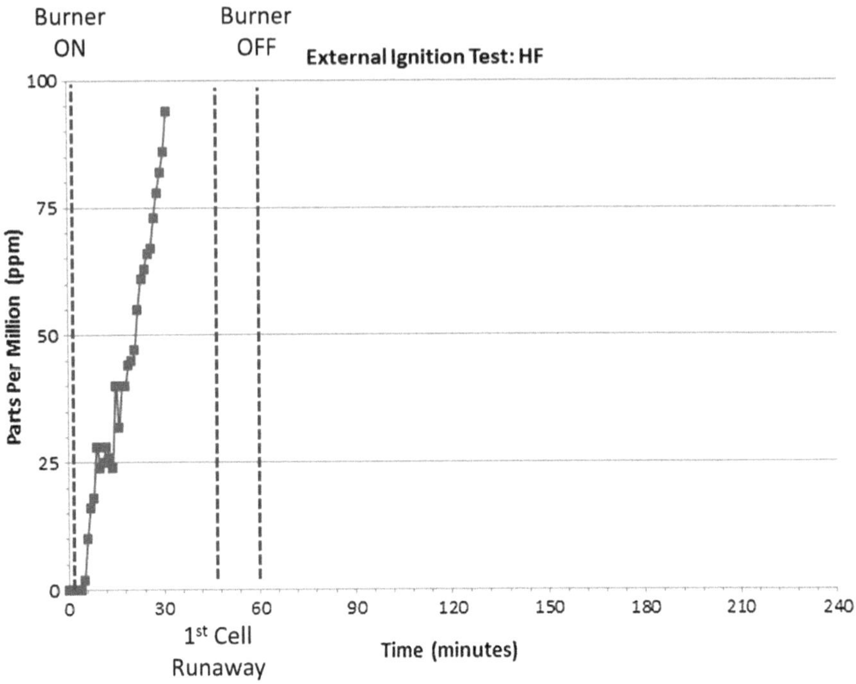

Fig. C.2 HF detected at the exhaust vent (detector maximum range was 100 ppm; all measurements after minute 30 were "over range," indicating the HF values were greater than 100 ppm for the remainder of the test

Appendix D
Internal Ignition Test: Temperature Plots

See Figs. D.1, D.2, D.3, D.4, and D.5.

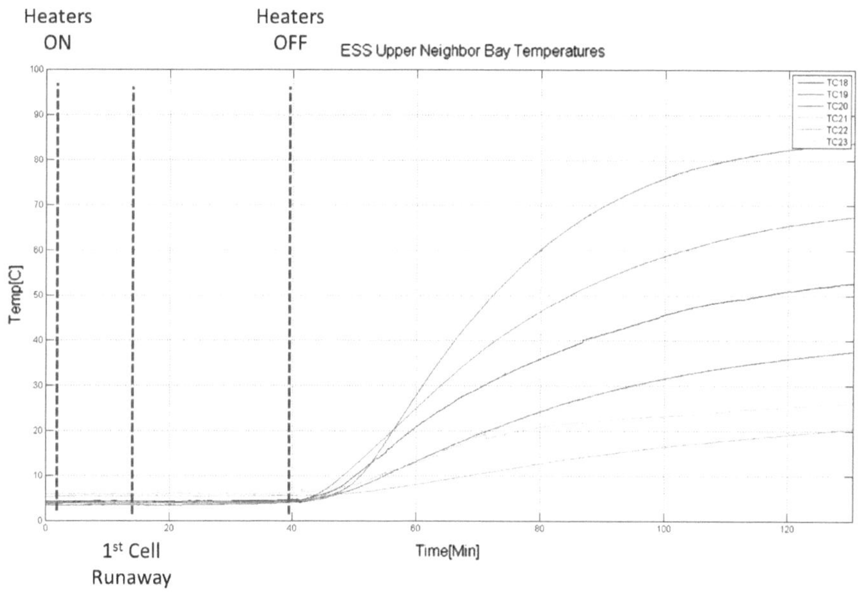

Fig. D.1 Pod 7 temperatures

A.F. Blum and R.T Long Jr., *Fire Hazard Assessment of Lithium Ion Battery Energy Storage Systems*, SpringerBriefs in Fire,
DOI 10.1007/978-1-4939-6556-4

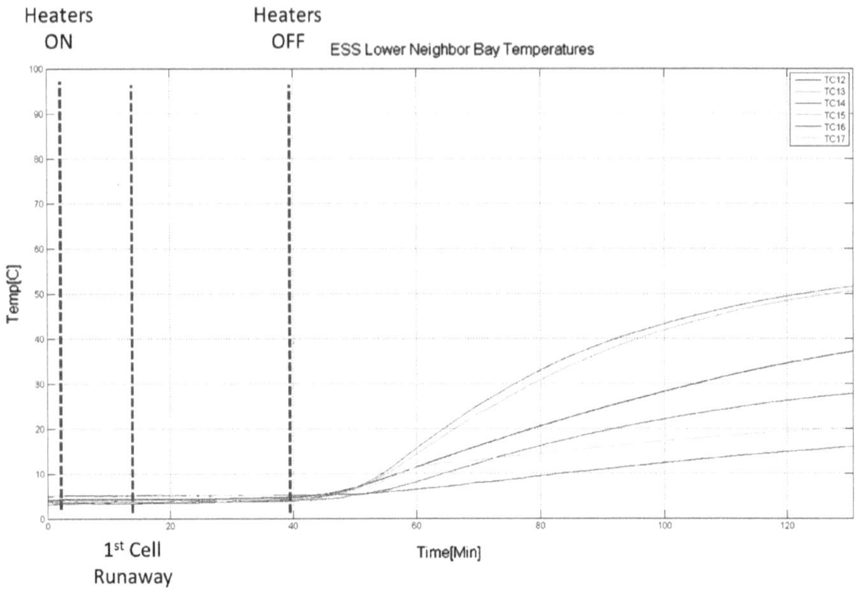

Fig. D.2 Pod 5 temperatures

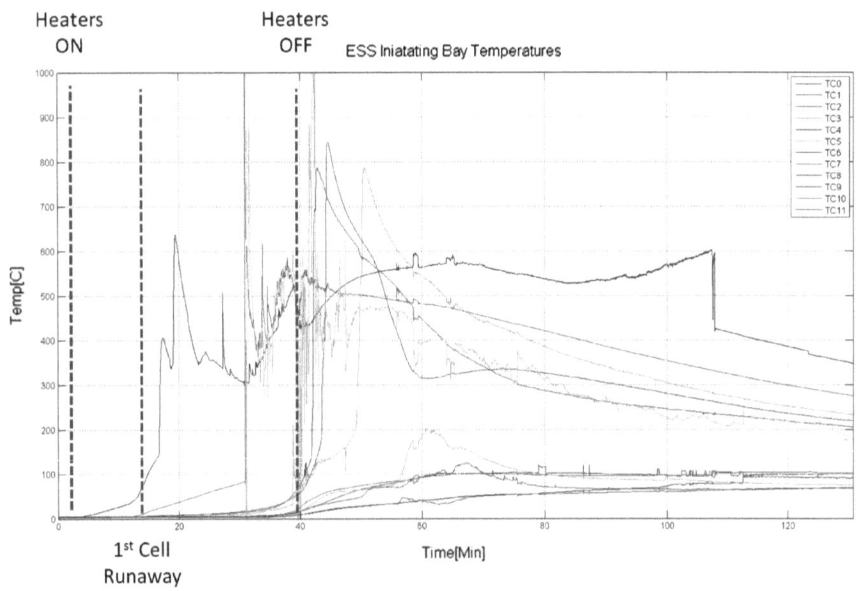

Fig. D.3 Pod 6 (initiator pod) temperatures (noise observed in the data is consistent with electrical interference that occurs during voltage leakage from the damaged batteries after thermal runaway)

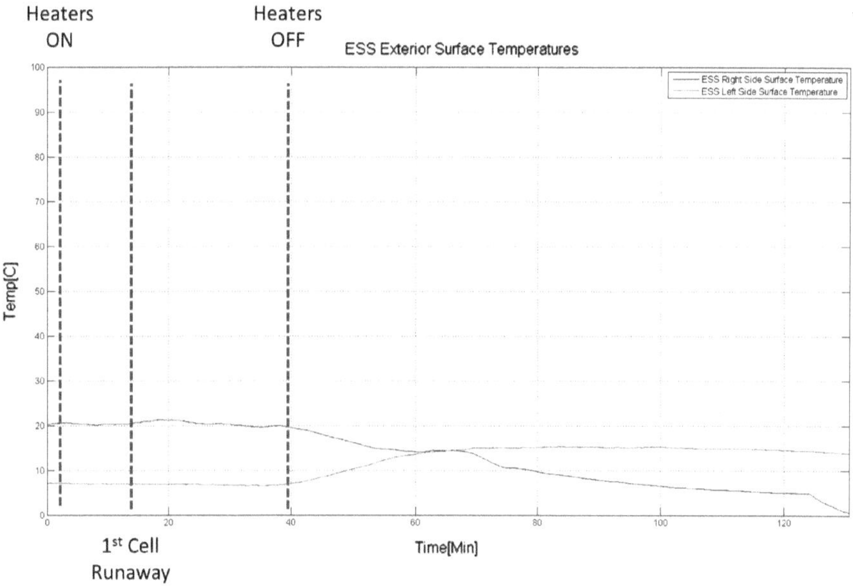

Fig. D.4 Exterior Powerpack surface temperatures

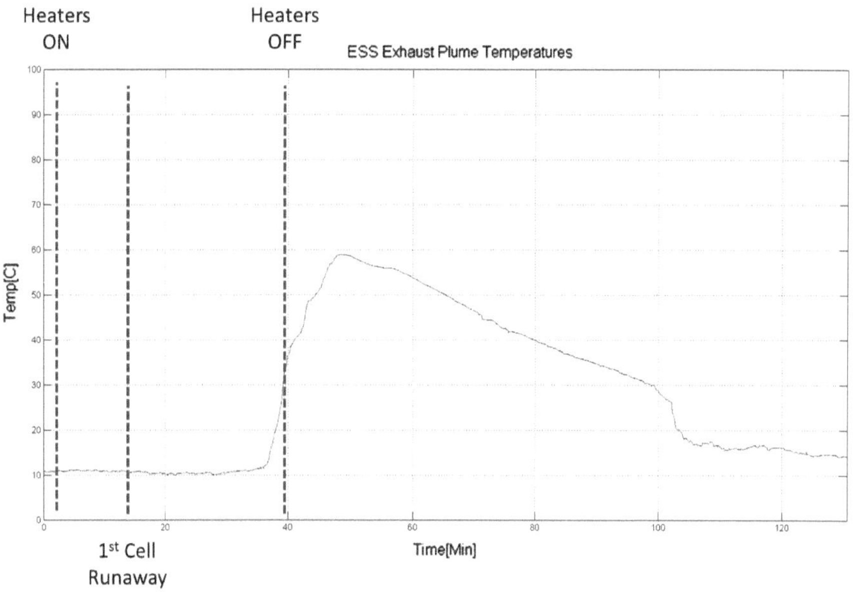

Fig. D.5 Exhaust vent temperature

Appendix E
Internal Ignition Test: Pressure Plots

See Figs. E.1, E.2, and E.3.

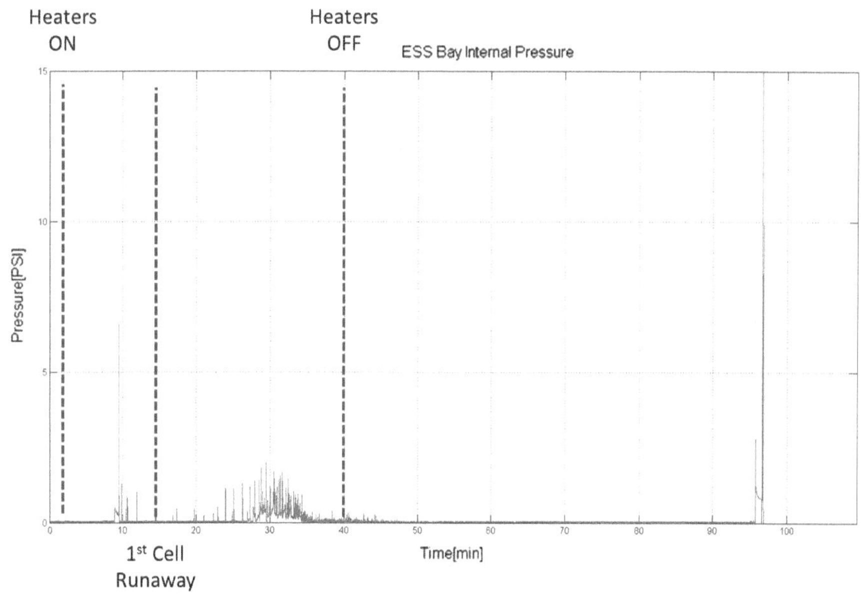

Fig. E.1 Pod 6 (initiator pod) pressure

A.F. Blum and R.T Long Jr., *Fire Hazard Assessment of Lithium Ion Battery Energy Storage Systems*, SpringerBriefs in Fire,
DOI 10.1007/978-1-4939-6556-4

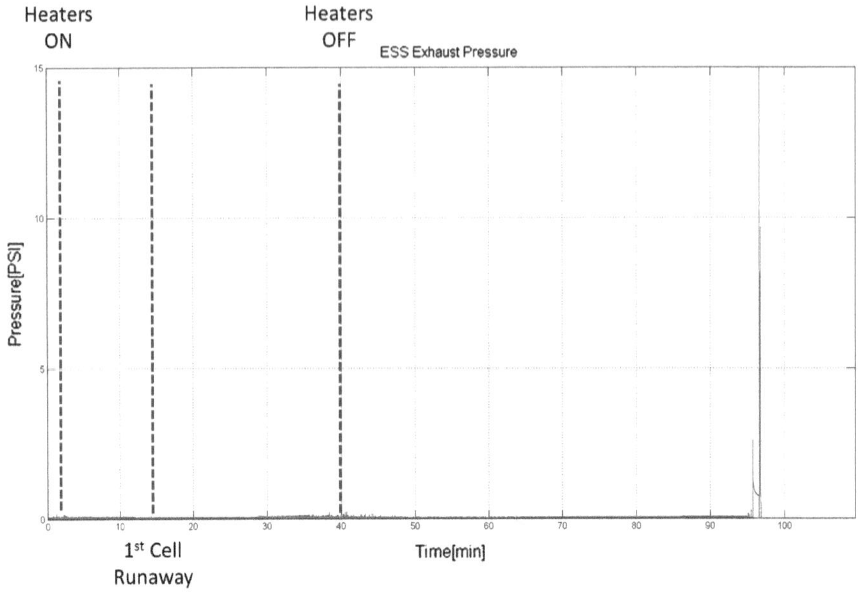

Fig. E.2 Powerpack exhaust manifold pressure

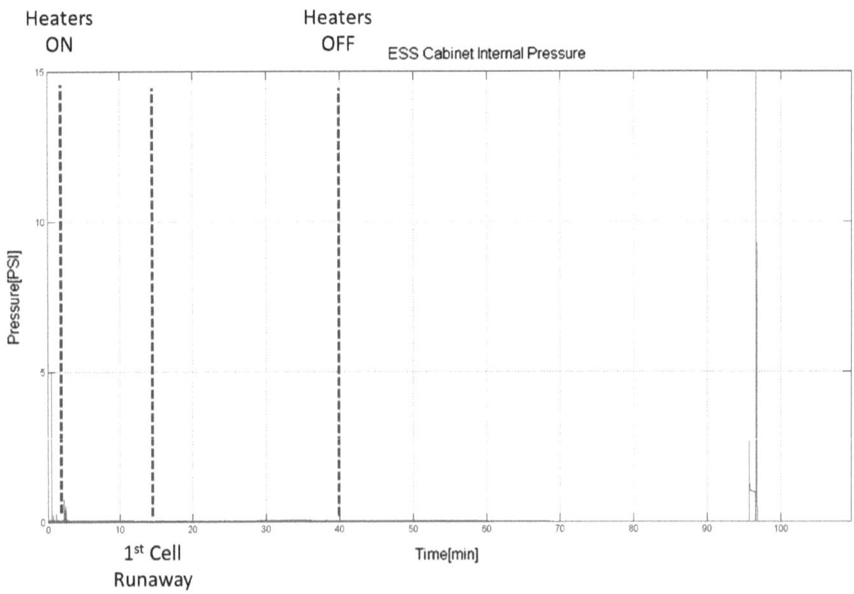

Fig. E.3 Powerpack cabinet pressure

Appendix F
Internal Ignition Test: Gas Sampling Plot

See Figs. F.1, F.2, and F.3.

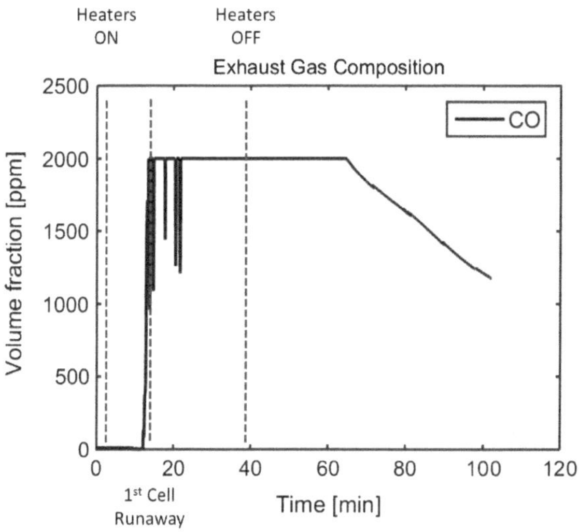

Fig. F.1 CO detected at exhaust vent

A.F. Blum and R.T Long Jr., *Fire Hazard Assessment of Lithium Ion Battery Energy Storage Systems*, SpringerBriefs in Fire,
DOI 10.1007/978-1-4939-6556-4

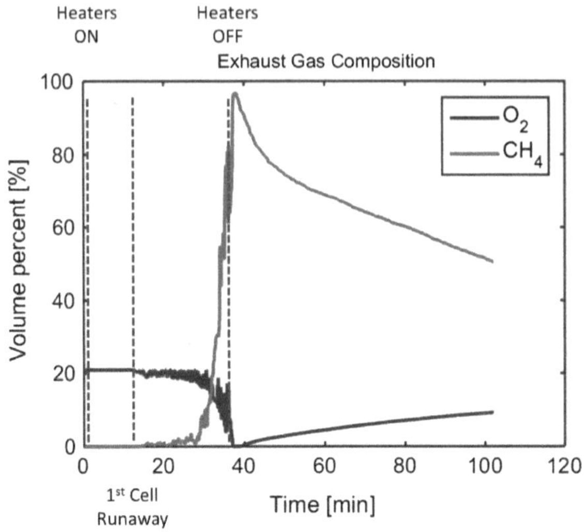

Fig. F.2 CH_4 detected at exhaust vent

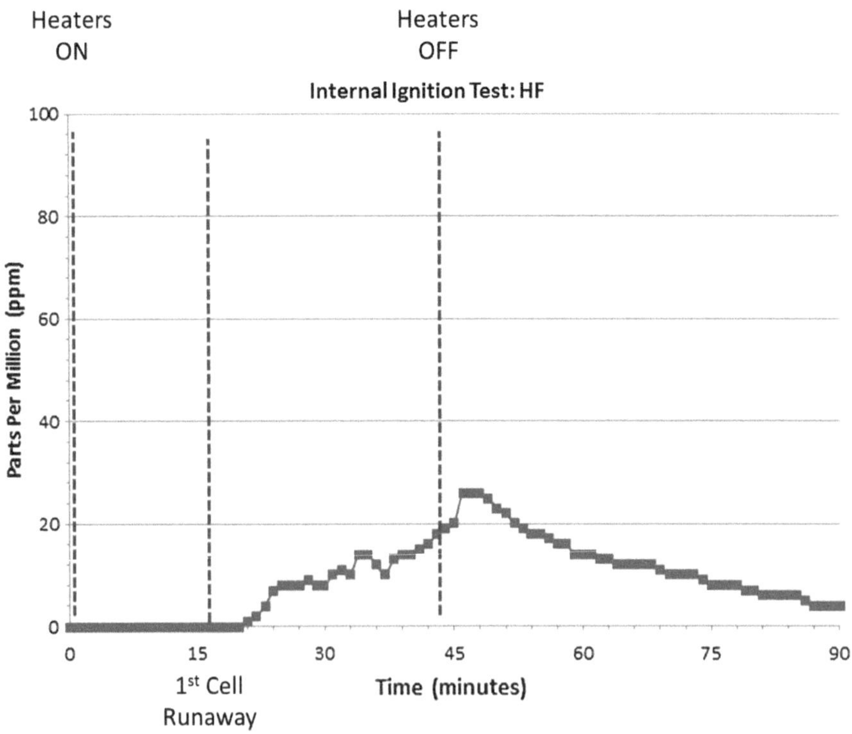

Fig. F.3 HF detected at the exhaust vent

FSC
www.fsc.org
MIX
Papier | Fördert gute Waldnutzung
FSC® C083411

Kolibri 360 GmbH
Ferdinand-Jühlke-Straße 7
99095 Erfurt, Deutschland
produktsicherheit@kolibri360.de